D1459792

Killing the Koala and Poisoning the Prairie

Killing the koala and
poisoning the prairie :
Australia, America, and
the environment
33305234698607
3an 12/24/15

Killing the Koala and Poisoning the Prairie

AUSTRALIA, AMERICA, AND THE ENVIRONMENT

Corey J. A. Bradshaw and Paul R. Ehrlich

The University of Chicago Press CHICAGO AND LONDON

Corey J. A. Bradshaw is the Sir Hubert Wilkins Chair of Climate Change in the Environment Institute at the University of Adelaide in South Australia.

Paul R. Ehrlich lives in California, where he is the Bing Professor of Population Studies and the president of the Center for Conservation Biology at Stanford University. He is the author or coauthor of many books, including, most recently, *Hope on Earth: A Conversation*, also published by the University of Chicago Press.

The University of Chicago Press, Chicago 60637
The University of Chicago Press, Ltd., London
© 2015 by Corey J. A. Bradshaw and Paul R. Ehrlich
All rights reserved. Published 2015.
Printed in the United States of America

24 23 22 21 20 19 18 17 16 15   1 2 3 4 5

ISBN-13: 978-0-226-31698-7 (paper)
ISBN-13: 978-0-226-27067-8 (e-book)
DOI: 10.7208/chicago/9780226270678.001.0001

Library of Congress Cataloging-in-Publication Data

Bradshaw, Corey J. A., author.
Killing the koala and poisoning the prairie: Australia, America, and the environment / Corey J. A. Bradshaw and Paul R. Ehrlich.
pages cm
Includes bibliographical references.
ISBN 978-0-226-31698-7 (cloth : alk. paper) — ISBN 978-0-226-27067-8 (e-book) 1. Environmental degradation—United States. 2. Environmental degradation—Australia. 3. Environmental protection—United States. 4. Environmental protection—Australia. 5. Ecology—Religious aspects. 6. Ecology—Political aspects. I. Ehrlich, Paul R., author. II. Title.
GE150.B73 2015
304.2'80973—dc23   2015014671

♾ This paper meets the requirements of ANSI/NISO Z39.48-1992 (Permanence of Paper).

For K. & little C., Anne & Lisa—
who make life worthwhile

Contents

(With hints on what to expect in each chapter.)

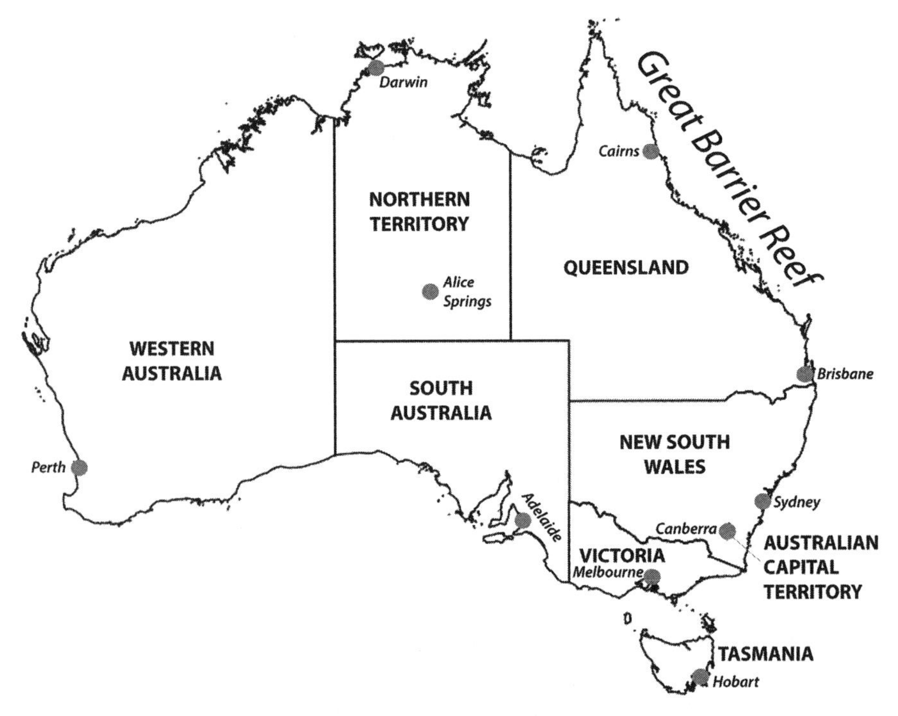

Map of Australian states and major cities.

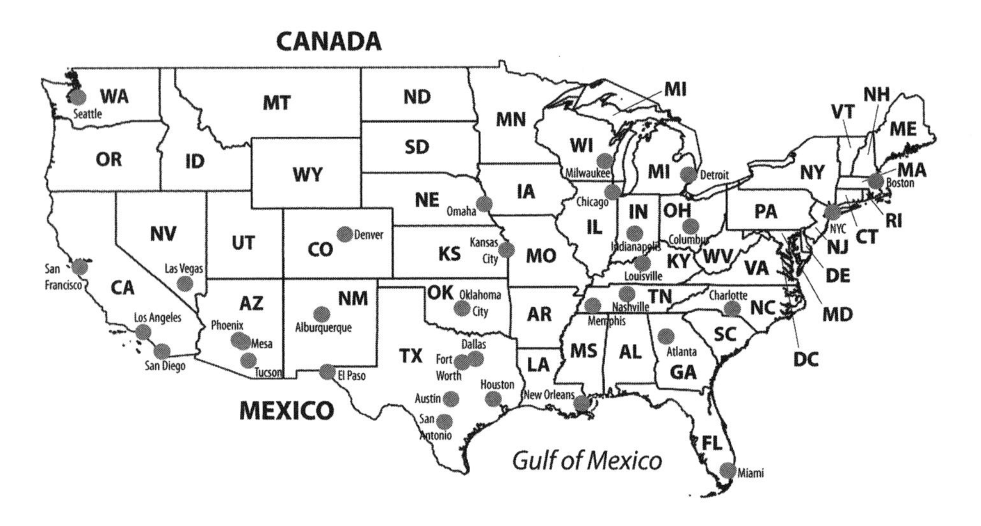

Map of United States and major cities mentioned in the text.

Preface

For much of the last century, Australia, like many other nations, looked at the United States from afar in a sort of confused awe. America's seemingly endless wealth, the can-do-anything-no-matter-how-crazy attitude, and its scientific and technological prowess and military power made little old Australia feel far away and inadequate. Another source of Australian fascination concerns US politics—no matter how much disdain Australians can generate for their own politicians, they always seem to cheer up a little when they see just how crazy American politics can get. For most Americans, Australia probably does not even rate on the foreign Richter scale (which unfortunately represents their attitude to almost every other nation), apart from some vague notion of bouncy, pouched animals, crocodile hunters, and endless deserts and beaches filled with countless dangerous, pointy, and poisonous creatures. Australians, on the other hand, have been spoon-fed a large amount of pop-culture drivel from American television and movies, and so most have at least a superficial knowledge of what appears to make—or fails to make—the United States tick.

In reality, the relationship between the two countries is profound on many cultural, political, and military levels. Australia and the United States bonded during the Second World War; for the first two years of the war, the Australian Army fought in Greece, Crete, and the western desert. The "diggers," as Aussie troops are known, were heavily engaged fighting Nazis when the Japanese struck southward in 1942. Australia seemed very much exposed, as the Japanese triumphed seemingly everywhere, even capturing the British Empire's "eastern Gibraltar" at Singapore (which had welcomed units of the Royal Australian Air Force a year before[1]) in mid-February 1942 in a brilliant and brutal campaign. But the American strategic naval victory at the Battle of the Coral Sea on May 4–8 blunted the Japanese advance and

P.R.E. in Botswana. African elephants are in deep trouble because of the ivory trade, but this one in the Okavango delta of Botswana just seems curious about biologists. Photo by John Schroeder.

kept the sea lanes to the east open, thus allowing US troops and supplies to reach Australia. Like the United States, Australia has only been directly assaulted once since 1900—also by the Japanese—with an aerial bombing campaign in 1942 on its most northern city, Darwin.

The Americans were generally welcomed to Australia with open arms as a bulwark against invasion, despite occasional violent disagreements over who should be dating Australian women. A very friendly attitude of Australians toward Americans persisted—it was obvious to Paul (henceforth P.R.E.) and his wife and collaborator, Anne, when they spent a delightful sabbatical year in Sydney in 1965–66 that led to lifelong friendships and many returns to Australia. Today Australia still looks to the US military to protect its interests in Southeast Asia,[2] much to the chagrin of Australia's major trading partner, China. Any Australians who have traveled to the United States have probably experienced a similar sense of camaraderie and acceptance, especially when speaking, apart from being asked incessantly to repeat their requests.

When we met, we discovered immediately that beyond sharing a

C.J.A.B. on Macquarie Island in the Southern Ocean. C.J.A.B. once studied elephant seal populations and was based at Macquarie Island station for four years. This seven-year-old female is fast asleep from anesthetics for measurement. Photo by Michele Thums.

deep appreciation of good wine and erudite company, we also were fascinated by the similarities and differences of our two nations (Paul, United States; Corey, Australia), which happened to be the two nations we both knew best. We have observed the similarities and differences of Australia and the United States firsthand. We see that our countries are friends, but that they do not necessarily learn from each other's mistakes—in fact, recent history suggests that we do exactly the opposite and parrot each other's failures. The eyes we see through are trained as those of environmental scientists and evolutionary biologists, and we both have beautiful wives (only one each), brilliant daughters (again, only one each), and the reputations of being gentle souls who strive never to express a strong opinion or give offense. If you believe that last clause, you should put this book down and go have a glass of wine instead. In a series of conversations enlivened and lubricated by some lovely wine at Ngeringa Vineyards[3] in South Australia, we came up with the idea of comparing and contrasting the environmental pasts and futures of our homelands, something that deeply interested, concerned, and stimulated us both. Then it occurred to us that we could combine that enterprise with an opportunity to tell people the unvarnished *Truth*—in other words, our personal opinions backed up by a substantial body of scientific evidence—about the human predicament: the perfect storm of problems confronting civilization. That was the genesis of this book.

Prologue

Do you believe that our universe containing some hundreds of billions of galaxies made of perhaps 300,000,000,000,000,000,000,000 (300 sextillion) stars was created some 6,000 years ago by a supernatural entity that retains a strong interest in your sex life? Do you feel that people with dark skin are genetically inferior to those with beige skin? Do you think that scientists are perpetrating a global-scale climate-change hoax? Do you agree with the *Wall Street Journal* economists who think that the human population and the economy can grow forever? Do you idolize Rupert Murdoch for his pursuit of truth in the media? Do you believe that women should be submissive to men? Do you think that environmentalism is a greenie, communist plot to bring down capitalism? Do you feel that worrying about species' extinctions is an unnecessary distraction from society's "real" problems? Did you learn in school that the Boston Tea Party was "anti-government"? Do you think that Australia is a small country in Europe?

If so, you should buy this book, but be prepared to become skeptical about many of your most cherished ideas. At first you might just be enraged, because it has been written by two scientists who are sufficiently furious at the state of our global environment and society to forget about political correctness. We are ecologists who are willing, even eager, to disagree with you on all the points listed above, and we unashamedly attempt to recruit you into the growing mass of people who are determined to divert society from its "business as usual" path toward disaster. We are friends who both know and love Australia and the United States, and feel like jilted lovers (well, not to each other). We are fed up—nay, disgusted—with the way politicians and the press ignore the realities that civilization is sliding toward irreversible damage, that universities are not providing any leadership to change our course to destruction, and that too many of our academic colleagues are busy doing more and more sophisticated studies of more and more trivial problems (sorry, friends). We are tired of the erosion of public education in both nations (which are ironically best positioned to get their environmental acts together), overlooked or encouraged by politicians who would never be elected by a public that had a basic understanding of how the rest of the world really works. We tend to agree with our colleague ecologist Harry Recher that the roots of decay in our educational system are deep. As he said:

> Any society whose culture is focused on mall crawling and sports will never have a decent educational system.

We are especially angry over the increasing concentration of wealth and power in the hands of anti-thought plutocrats within nations, and with the purveyors of what Naomi Klein has called "disaster capitalism"[1] on an international scene still plagued by imperialist resource wars and crazy religious conflicts, and lacking any sensible plan of global governance to deal with global problems. We are incensed about what all of this is doing to the world that today's youngsters—our own descendants and yours—are going to inherit.

So hang in there if you would like to hear what two scientists (and many of our colleagues) really think, and what we believe the world can learn from a comparison of the different ways that two very similar cultures could save our societies, rather than being allies (as we currently are) in a war to destroy the environment on which we all depend. We promise not to hold anything back, and if you want to give this book a fair read, you will consult some of the references to appreciate the massive evidence that stands behind our strongly held opinions. If in the end you still hold the views listed in the beginning of this prologue, please donate the book to your local public library or high school.

1 Ausmerica

Americans can do anything.

JULIA GILLARD, former Prime Minister of Australia,
addressing the US Congress in 2011

Trees cause more pollution than automobiles do.

RONALD REAGAN (1981),
former President of the United States

Man and the environment are meant for each other.

TONY ABBOTT (2014),
Prime Minister of Australia

I know the human being and fish can co-exist peacefully.

GEORGE W. BUSH (2000),
former President of the United States

> Paul, you've been tackling the plutocrats, science denialists, and evidence-free ideologists for longer than anyone I know. Does it make you angry or frustrated (or both) that it seems to be getting worse and not better?

I (Corey Bradshaw, henceforth C.J.A.B.) remember asking Paul that question in 2009 as we ate lunch on the Stanford University campus near San Francisco, California. We had just met after being introduced by a mutual colleague, Professor Gretchen Daily, also of Stanford University. I was visiting the university for the first time and had been invited by Gretchen to give a presentation of my research to her lab group. Paul and I had hit it off immediately.

There are few scientists more famous in my field than Paul Ehrlich—he has a long and celebrated career of top-notch science, has written dozens of popular books, and is, of course, most famous for his first book, *The Population Bomb* (1968),[1] which triggered decades of debate about human overpopulation. I was understandably both nervous and elated to be in the presence of such scientific royalty. I cannot recall the specifics of his response to my question—it had something to do with keeping your sense of humor and maintaining your internal environment with good wine—but the man's vibrant, cutting intellect, sense of humor, and emotively filled diatribe (this book will give you a comprehensive overview of that subject) impressed me immediately. This was a scientist nearly two generations my senior who could fire off passionate and erudite arguments with military precision. I had to get to know him better.

I (Paul Ehrlich, henceforth P.R.E) rarely run into youths who have the good sense to agree with me on virtually everything. How could such a young* man have figured out how the world works with relatively so little experience with it? I'm part of a gang of old farts who are desperately trying to change the world that our grandchildren are facing, and here was a brilliant ecologist barely out of his academic diapers concerned about the fate of his beautiful three-year-old daughter.

Several dinners, bottles of wine, and deep conversations later in the presence of our families, we decided to write this book. We took it in turn to visit each other's country, so the subject material grew as we identified more and more reasons why Australians and Americans have a lot to learn from each other's mistakes. There is an almost spooky similarity in the environmental and political problems both of our countries are experiencing, even though the origins of them are often utterly divergent.

Yes, there are many similarities between Australia and the United States. Just think about the rough dimensions of our two countries: both nations are large, with Australia just slightly smaller than the forty-eight contiguous states of the United States. In fact, Australia is the world's sixth-largest country, covering 7.69 million square kilometers (2.97 million square miles). The entire United States, with its fifty states,

*Age is relative, of course. C.J.A.B. was thirty-nine when he met P.R.E. for the first time.

is the world's fourth-largest country (or third, depending on what China is currently claiming), covering 9.83 million square kilometers (3.79 million square miles).[2] Both cultures are derived originally and principally from what is now the United Kingdom, despite being today a mix of hundreds of other cultures. English, or what approximates for English in each, is the dominant language, but with rapidly rising language minorities, such as Spanish in the United States and Mandarin Chinese in Australia. Both are examples of super-consuming, overdeveloped, rich, literate countries, but with two Americans consuming the same amount of resources as three Australians. But both have footprints more than threefold that of Costa Rica or Chile. Australia and the United States each spend roughly the same portion of their gross domestic product on education, and both have (by standard definition *only*) highly educated populations. They are among the leaders in environmental sciences, along with the United Kingdom and some European (especially Scandinavian) nations and Japan, with China moving up fast. Both are top greenhouse gas emitters, with each country producing today about fifteen times as much per person as does India.

We think Aussies and Yanks (to use Australian parlance) generally get along well for a number of reasons. Both countries have a frontier spirit; both threw off the British yoke (although Australia keeps some of the royal regalia around for obscure and sentimental reasons); people in both nations tend to be direct and enjoy the out-of-doors and off-color jokes. Today there are over 60,000 US-born people living in Australia,* mostly in Sydney and Melbourne, and about 90,000 Australians in the United States. In 2011, 460,000 Americans visited Australia, and 1.2 million Australians visited the United States. These affinities and our strangely convergent personalities are probably why we two hit it off immediately. But above all, we do not like pussyfooting around.

Of course, there are some big differences between Australia and the United States too—many of environmental significance that we discuss in detail in this book. We are not just talking about their wildly different floras and faunas. Although comparable in size, the United States is about ten times as densely populated as Australia. The United States might be ahead in income per person, but the continuing of Ron-

*According to 2006 census data.

ald Reagan's "Hood Robin" program of redistributing wealth from poor to rich has put the United States well behind Australia in fairness of wealth distribution. Australia's more fragile environment is indicated in many ways, such as Australia having the highest number of recently extinct mammals compared to every other country in the world. Both nations had large indigenous populations and treated them (and still largely treat them) very badly, but the US record was probably worse—although ask Australian Aborigines, and they might have cause to disagree with that assessment. Both countries are nations of immigrants, although the timing and composition of immigrant waves were radically different. Many early European immigrants to both countries were convicts, but both are now very efficient at growing their own. Indeed, there are twice as many people in prisons than living on farms in the United States.[3] In Australia there are closer to nine times as many people living on farms as are incarcerated,[4] although in 2014 the latter number hit an all-time high[5] of 33,000. The United States is also way ahead in the department of gun violence, in part due to an insane lack of controls on guns and ammunition (based on a silly misreading of the US Constitution). The United States also has a Supreme Court justice who has seriously considered[6] the notion that your next-door neighbor carrying shoulder-launched, anti-aircraft missiles should be constitutionally protected under the United States' infamous "right to bear arms" Second Amendment.[7] The bottom line is that the United States has roughly five times the proportion of people in prisons compared to Australia, but a larger proportion of Australians are victimized by crime because relatively more Australians live in big cities compared to citizens in the United States.[8]

Disturbing trends shared by the two nations are increasing right-wing, corporate control of much of the mainstream news media, with incredible power vested in the hands of the Murdochracy, that gang of morally questionable individuals producing "news" sources like *Fox News*, the *Wall Street Journal*, and the *Australian*. These promote the trend toward plutocracy and theocracy in both nations, where a politician is toast (hold the vegemite) if he or she does not at least pretend to believe in supernatural entities—the obvious exception being former Australian Prime Minister Julia Gillard, who is an atheist and resided at the Lodge (the official residence—analogue to the White House) with her unmarried male partner. Neither nation deserves the major blame for the Murdoch stain on the human race—he was born in Australia and naturalized in the United States, and buys influence for the greedy of both nations.[9]

Despite their reputations for technical know-how and a can-do spirit, the world that supports Yanks and Aussies is obviously going down the drain at an alarming rate, and even many of our colleagues appear content to let it happen. We often see the media portray so-called "debates"—especially with respect to climate disruption—between the evidence-wielding majority of scientists versus the data-free opinions of the corporate-backed and ideologically crazed "denialist" pundits. But there is no "debate"—it is a bare-knuckle street fight, and when many more scientists appreciate this unfortunate reality, the world will be much better off. The way we see it, when politicians, corporations, and special-interest lobbyists threaten our grandchildren, the gloves must come off. Fortunately for scientists like us, we have knuckles hardened by real data. To paraphrase Naomi Klein, in her critique of international capitalism, *The Shock Doctrine*,[10] "The business plan of the oil industry is to destroy the world." Ours must be to prevent them, and others of their ilk, from achieving that goal. We cannot allow Australia and America to remain close allies in the war on the environment.

The following chapters in this book weave a comparative story of our two countries—a story that we contend has important global implications. Chapters 2 and 3 focus on the rise of pre-European and post-European societies in the two countries, and how humans have irrevocably altered their natural environments. Chapter 4 explains why without intact ecosystems, we are overall poorer nations. Chapter 5 lays out the current state of that damage in both countries. Chapters 6 and 7 summarize the evidence for rising toxicity and the continuing pressure of overpopulation (yes, even in Australia!). Chapters 8 and 9 deal with the politics of environmental policy failure, driven mainly by greed and religious conservativism. Chapter 10 spells out clearly that we have almost no time to turn these backward policies around, and chapter 11 sets up a blueprint on what we must change to avoid the worst environmental and political crises.

The deep historical, cultural, economic, and military ties between our two countries should not be ignored, whether or not they are understood or appreciated by most of their citizens. Many Australians still aspire to emulate the successes of the United States, in politics, foreign policy, industry, business, technology, and academia, but fail to learn from the colossal mistakes that the United States has made in each one of these areas. Americans, on the other hand, would do well to adopt some of the great Australian approaches to managing society. But none of this is possible without the continued functioning of both countries' life-support systems—their natural environments. For too long their citizens have been biting the hand that feeds them, so it is high time we make some sweeping changes to our societies to fix the damage already done and to avoid the ensuing catastrophes that are increasingly imminent. Australia and America are great nations, but we are both highly susceptible to our own greed and stupidity. Let's change that.

2 Enter the Naked Ape

This is our land. It goes back, a long way back, into the Dreamtime, into the land of our Dreaming. We made our camp here, and now all that is left of our presence are the ashes and the bones of the dead animals the young men had killed. Soon even our footprints will be carried away by the wind.

S. M. COUPE, *Historic Australia*[1] (1982)

Whitefella comin'.

After the title of D. S. Trigger's book,[2] taken from Aborigines' warning to others of the impending arrival of European colonists

The most recent evidence[3] suggests that as much as 120,000 years ago, human beings started leaving Africa, *Homo sapiens'* evolutionary cradle, and began the war on the environment—the monumental task of screwing up the rest of the planet. Despite our ruthless efficiency at achieving that inadvertent objective, the history of human occupation in Australia and North America could not be more different. Yet the influence of those early people on the ecosystems of both regions ultimately converged. Humans have a remarkable capacity to adapt to pretty much any suboptimal condition that Earth's environment can throw at them, because our capacity for survival is embodied in our striking ability to evolve culturally and modify our surroundings to suit our needs. There is no better ecosystem-engineering species on the planet. Sadly, this efficiency also means that when enough of us get together, we can do a surprisingly large amount of damage to our own life-support systems. We are specialists at sawing off the proverbial limb upon which we are sitting.

We will start with Australia because humans arrived there long before they ever showed up in North America. The first Australians arrived about 50,000 years* ago from the Indonesian archipelago, making the Aboriginal ways of life some of the most ancient *in situ* (surviving) cultures on the planet. Indeed, outside of Oenpelli in the Northern Territory, P.R.E. and his wife, Anne, were thrilled, along with their friends the ecologists Andy Beattie and Chris Turnbull, to view an amazing display of rock art. A hundred meters or more of low rock underhang was decorated with paintings, often overlaid, some of which were said to date back 10,000 years or more. Yet the same day and a mile or so away, they could see the lineal descendants of the artists now doing their paintings on special paper imported from England. The early Aborigines also developed some important advances in stone tool-making and invented some of the earliest forms of open-water transport.[4] Those first people arrived by small vessels[5] across the Arafura and Timor Seas separating the Australian continent from the many large islands that today we call Indonesia, Timor-Leste, and New Guinea. They found a massive continent teeming with food. Wallabies, kangaroos, wombats, marsupial tigers, crocodiles, giant terror birds (*Genyornis*), barramundi, goannas, cockatoos, snakes, turtles, and many small marsupials made up the delectable smorgasbord that a talented hunter armed with a simple spear could acquire with relative ease. There were also many native plants such as yams, macadamia nuts, and various fruits that complemented the meat-rich diet. What a paradise those remote northern shores of Australia must have looked like to those first humans!

Yet it was far from a gentle land—northern Australia is to this day a region of intense extremes. Its weather system is infamous for its brutality to living things. Unlike the monsoons typical of the lower latitudes immediately to Australia's north, which are characterized by several "wet" seasons interspersed by periods of relatively little precipitation, northern Australia's rains come only once annually after an extended period of intense dryness. From C.J.A.B.'s personal experience, we must emphasize that the gradual and relentless in-

*Estimates vary between 40,000 and 60,000 years based on the dating precision of various artifacts found around the Australian continent. Most anthropologists accept the median estimate of 50,000 years as approximately correct, although even older occupation dates have been suggested.

Aboriginal rock paintings from central Arnhem Land near Kolarbidahdah Outstation. Photo by Corey J. A. Bradshaw.

crease in humidity and heat before those first rains arrive is almost enough to drive one insane. The brief ensuing wet season lasts a mere two to three months, with often no appreciable rain falling during the remainder of the year. This boom-and-bust cycle of available water means that all life—humans included—had to evolve or adapt quickly to survive this astonishing variation in the availability of life-giving water.

Not only is the extreme monsoon of the north a burden, some better-known aspects of Australian climate and geography also conspire to challenge life. Australia is the most water-stressed of the inhabited continents. Jared Diamond unflatteringly described Australia as the "driest, smallest, flattest, most infertile, climatically most unpredictable, and biologically most impoverished" continent on Earth.[6] The characteristic red dust, scrubby vegetation, and peculiar rock formations, so often reproduced on tourist pamphlets, make up a substantial proportion of the country's land area—approximately 40% of its 7.69 million square kilometers (2.97 million square miles)—is considered arid shrubland or desert. Most of these arid lands sit squarely

The arid Outback: what most of Australia looks like—little prospect for agriculture. Photo by Corey J. A. Bradshaw.

in the middle of the continent-country, with only the coastal fringes having climate patterns conducive to forest growth. But even the more vegetation-friendly parts of the country are subject to intense variation from year to year. The climatic phenomenon known as the El Niño–Southern Oscillation (ENSO) can bring heavy rains, floods, and cold fronts one year, and extreme droughts the next, with little apparent predictability. The last few years of record high temperatures, intense heat waves, and disastrous floods in Australia are not only testament to its innate climate variability, these phenomena are probably getting worse as the planet warms. Including savannas* with tree can-

*A mixture of grassland and trees.

opy cover greater than 20%, Australia had a mere 30% of its total area covered by forests when Europeans arrived some 200 years ago. Since then, while forests have dwindled to cover less 20% of the country,[7] the total area of desert has changed little.

Australia is also an old continent. Barring a few localized exceptions, it has not experienced any major volcanic activity over most of its surface* for many millions of years. Given the low rainfall over most of the country, there are few areas that have received substantial siltation from flooding rivers. Nor has Australia had any major glaciation† over the last few hundred million years.[8] The lack of volcanic activity, siltation, and surface scouring means that most soils in Australia have not been replenished for millions of years. As young soils (recently erupted rock broken down by erosion) age, their nutrients can be leached by the processes of plant growth, water movement, and fire, so, generally speaking, the longer soils go without volcanic or other disturbances, the more depleted they are of nutrients essential for plant growth.[9] Of course, certain plant species have evolved to cope with nutrient-poor soils, of which the wonderfully diverse eucalyptus (gum) trees (including the genera *Eucalyptus*, *Corymbia*, and *Angophora* in the family Myrtaceae) are prime examples, but the vegetation growing on such soils is usually not conducive to domestication and mass cultivation. In fact, there are few native Australian species that have been forced into cultivated servitude even today—the macadamia‡ is an exception.

It is little wonder, then, that for the entire 50,000-year(ish) history of Australian Aborigines, not one of the hundreds of different tribes developed any substantial agriculture or livestock-rearing capacity. There were, however, some sedentary "farming" settlements in southern Australia where fish-trapping was employed.[10] There is also evidence of stored "grains" for baking and trading[11] (seeds from the plant genera *Panicum*, *Trigonella*, and *Marsilea*) found in some larger Ab-

*However, there are several regions in Australia that have experienced volcanic eruptions over the last 20,000 years, such as the Mount Gambier region of South Australia, Western Victoria, and the Atherton Tablelands of North Queensland.

†There were, however, a few glaciers in the Late Pleistocene in the southeastern highlands and Tasmania.

‡However, Australia was not the first country to benefit from a large commercial cultivation of this native food. The United States (Hawai'i) was.

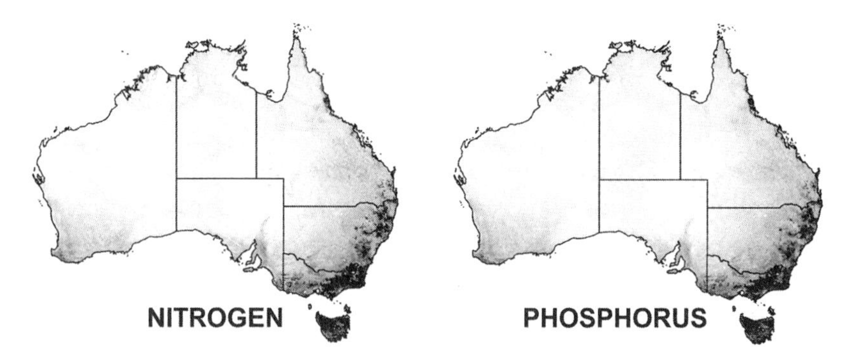

Mean total nitrogen (*left*) and phosphorus (*right*) for plant growth in Australia (blacker pixels indicate higher nitrogen or phosphorus). Most of the continent is characterized by extremely poor soils for plant growth. Data provided courtesy of Michael Raupach of the Australian National University and Peter Briggs of CSIRO. From M. R. Raupach, J. M. Kirby, D. J. Barrett, and P. R. Briggs, *Balances of Water, Carbon, Nitrogen and Phosphorus in Australian Landscapes: (1) Project Description and Results* (Canberra: CSIRO Land and Water, 2001).

original settlements, which might point to some basic agricultural capacity. However, the otherwise near-total lack of agriculture is remarkable because it is the only known case in the world where such a long-lived culture sustained itself almost entirely via hunting and gathering. Even with substantial trade established with the Makassan culture of southwestern Sulawesi (Indonesia) some hundred years prior to European settlement, the Aboriginal tribes of northern Australia still demonstrated almost no agrarian capacity. Apart from some small-scale production of various bush-tucker* crops and cattle-rearing ventures, Australian Aborigines are even today largely non-agrarian. Nonetheless, Australian Aboriginal groups have been among the ultimate sustainable societies.

The development of agriculture, allowing one family to produce more food than it needs, has always been at the core of humanity's rise to planetary dominance. From the expansion of modern human farmers out of the Fertile Crescent of the Middle East to the establishment of the ancient Chinese empires, an abundance of food from ever-larger-scale agriculture was essential to allow more and more specialization, which has been the basis for our societal development.

*Literally, "food from the forest."

Agriculture allowed many to give up procuring food and become warriors, priests, engineers, administrators, and the like. However, agricultural expansion is proving to be humanity's Achilles' heel because it is the principal driver of deforestation and extinctions around the planet. Indeed, over the last fifty years, agriculture has been identified as the main reason for the destruction and degradation of tropical forests,[12] home to most of the world's species.[13] Equally important, agriculture is a major emitter of greenhouse gases and, thus, a driver of climate change, which is in turn a prime threat to the productivity of the agricultural system itself.

One could assume, quite logically, that Australian Aborigines would have had little impact on the ecosystems in which they lived. A combination of low population density, lack of an extensive technological tool kit beyond simple stone implements, and no appreciable agriculture would suggest little capacity for major ecosystem engineering. Nothing could be further from the truth, however. Australian Aborigines were, in fact, master engineers of their environment, quickly depleting many medium- and large-bodied species from the landscape probably within only a few thousand years of their arrival.[14] Species such as the giant diprotodontid "wombats," weighing up to 1,000 kilograms (2200 pounds); the flightless bird titan *Genyornis*, which towered over the modern-day emu and weighed up to 100 kilograms (220 pounds); tree kangaroos of the genus *Dendrolagus*; several large kangaroos, including the giant *Procoptodon goliah*, which stood 2 meters (6 feet 7 inches) tall and weighed over 250 kilograms (550 pounds); and the short-faced kangaroo *Sthenurus*, which was up to twice as large as modern kangaroos; giant 100-kilogram (220-pound) *Protemnodon* wallabies; the 500-kilogram (1,100-pound) marsupial "hippo" *Zygomaturus*; as well as *Dorcopsis*, hare, rock and nail-tail wallabies, pademelons, various other kangaroos, many true wombats, potoroos, echidnas, cuscus, marsupial "tapirs" of the genus *Palorchestes*, and the carnivorous marsupial "lion" *Thylacoleo*—all perished about the time or shortly after human beings arrived in Australia.

While controversies still surround some of the explanations behind the megafauna disappearances in Australia and elsewhere, there is mounting evidence that humans were the principal drivers of most, if not all, of these extinctions. It is entirely plausible that climate shifts during the last 50,000 years, including the rapid warming of about

Some of the beautiful, if weird, megafauna that went extinct shortly after the Aborigines made it to Australia. *Upper: Diprotodon; middle left: Protemnodon anak* (giant wallaby); *middle right: Sthenurus* (short-faced kangaroo); *bottom: Genyornis* (*left*) with a modern emu (*right*) for comparison. All line drawings copyright Peter Murray. Courtesy of Chris Johnson, University of Tasmania.

5 degrees Celsius (9 degrees Fahrenheit) defining the beginning of the Holocene geological period 12,000 years ago, contributed to their demise. However, both direct evidence from archaeological finds[15] and indirect insight from computer simulations[16] demonstrate the ability of small populations of Stone Age hunter-gatherers to eradicate even the largest species in as little as a few hundred years.

But the Aborigines' influence was not restricted to efficient hunting; they were extraordinarily clever with their manipulation of fire on sometimes massive scales.[17] By the careful spreading of fire using burning sticks during various periods of the dry season, savanna- and forest-dwelling tribes were able to modify the vegetation structure to facilitate hunting efficiency. As for many ecosystems in North America, fires are a natural component of most forests in Australia, with huge areas burning at times during the dry season. With a little pyrotechnical encouragement, ancient Australians quickly learned that they could concentrate the species they hunted into areas that were easier to access, and they could remove much of the understory vegetation to increase the ease with which they could move through the bush and the distance they could spot potential prey. Controlled burning also had the advantage of attracting prey like kangaroos to the post-fire flush of succulent grasses. Aborigines became "fire-stick farmers" because they transported fire to wherever they wished and used it as a tool to create a landscape that was more to their benefit. As such, the Australian bush was far from "natural" by the time Europeans arrived, so much so that some have even suggested that prior to the proliferation of Aboriginal cultures, there was vastly more forest cover in the interior of the continent than what the Europeans first saw.[18] While the changes that Aborigines imparted to their landscape were substantial, it is important to remember that they pale in comparison to the massive loss of biowealth* caused by European agriculture. The bottom line is that humans, no matter their culture, have a regrettable and unparalleled ability to alter the landscape because of their intelligence, creativeness, and adaptability.

This culture of fire-stick farming was passed from generation to generation, until it eventually reached the new European settlers

*In relation to human well-being, all of biodiversity can be considered under the umbrella of *biowealth* (see chapter 4).

arriving in the late eighteenth century. These new arrivals quickly learned the advantages of this method and adopted it as their own. Of course, over time fire-stick farming evolved into a much more efficient practice through the use of kerosene-fueled drip-torches, the construction of roads into previously inaccessible areas, and, eventually, the employment of helicopters to cover much larger areas. Detailed research has shown that the patterns of bushfire in modern Australia are vastly different to what the first Australians were able to create, such that the structure of modern forests is also as different.[19] In northern Australia at least, there is now a firmly entrenched lust for fire management, with few areas escaping at least some burning in any given year. Despite a largely intact forest covering much of this region, the frequent burning has been hard on many small mammals and reptiles, with recent declines in numbers of up to 95% for several species reported throughout Western Australia, the Northern Territory, and Queensland.[20] We will delve more into the patterns of modern extinctions in a later chapter.

In contrast to the Aborigines' arrival in Australia some 50,000 years ago, human beings arrived in North America only about 13,000–18,000 years ago.* Like the true date of arrival of the first Australians, the debate surrounding the history of the first North Americans—how many different groups arrived and when—is somewhat contentious; regardless, the environmental impact of what has become known as the Clovis culture (because artifacts of this civilization were uncovered in the 1930s near the town of Clovis, New Mexico) is much less controversial. Penetrating the last extreme-climate barriers of northern Eurasia's extensive tundra zone and spreading across the 1,600-kilometer (1,000-mile) ice-free land bridge where the salt waters of the Bering Strait now separate Alaska and Russia, humans quickly (at least in a geological context—over a few centuries) populated much of North and South America.

Why did those ancient hunter-gatherers brave the inhospitable northern lands during a time when even clothing technology was still poorly developed? The never-ending quest for food, of course. Indeed,

*Although recent evidence now suggests a much older date. See http://www.nytimes.com/2014/03/28/world/americas/discoveries-challenge-beliefs-on-humans-arrival-in-the-americas.html.

the first known settlements on the far northeastern shores of Siberia were veritable mammoth graveyards—substantial evidence from these and Alaskan sites clearly shows that people were following this now-extinct species and other large animals typical of these northern lands at the end of the Pleistocene.[21] In these newfound expanses of abundant food came technological advances that sealed the fate of many medium- and large-bodied species, much as the Aborigines had done 40,000 years earlier to species in Australia. The so-called Clovis spearhead was a simple, yet amazingly sophisticated turn in the development of stone weaponry that ultimately spelled the end for many of North America's remarkable late Pleistocene beasts.

About the same time that human beings arrived in North America, many other species disappeared; in fact, over 70% of species weighing more than 45 kilograms (100 pounds) went extinct during this time. The saber-toothed cat, giant beavers, mammoths, stag moose, giant sloths, tapirs, peccaries, many species of horses, and camels (which, contrary perhaps to most people's intuition, evolved in North America) were all killed off. So, too, did short-faced bears, mammoths, and mastodons all disappear about the time that people discovered the bountiful continents of North and South America.

With the separation in time of 40,000 years, a completely different set of environmental conditions and constraints, tropical versus tundra hunting grounds, different tool kits, vastly divergent species, and altogether different historical trajectories, it is surprising that the causes of the megafaunal extinctions in Australia and North America are similar. Many scientists originally supported an entirely climate-driven cause for the loss of North American megafauna, especially considering that one of the greatest and most rapid shifts in climate occurred at the conclusion of the Pleistocene around 12,000 years ago and overlapped considerably with the last occurrence of many of the species described above. That said, there are more and more decidedly curious coincidences of large-bodied animals disappearing at about the time or shortly after humans arrived in different regions of the Americas. This is not to say that the two hypotheses are mutually exclusive—it is entirely consistent with observations that climate and hunting can work in synergy to increase extinction risk.[22] For North American megafauna in particular, it is likely that humans were the straw that broke the camel's already rather climatically strained back.

Human-caused megafauna extinction in North America. Upper left: mammoth (*Mammuthus primigenius*). *Upper right:* giant sloth (*Megatherium*); courtesy of Mauro Galetti. *Lower left:* mastodon (*Mammut americanum*); *lower right:* saber-toothed cat (*Smilodon*).

Food availability is where the first peoples of the Americas differed substantially from their Aboriginal counterparts. As we have discussed and for many quite understandable reasons, Australian Aborigines were largely bereft of agriculture, and while animal protein was at least at times abundant in Australia, the overall productivity of the landscape was low. In stark contrast, the first North Americans developed rather impressive societies built on both the eventual mastery of complex agriculture as well as the bounty of the land. In addition to being considerably more diverse than Australia in its regional climates—from the extreme cold of the Arctic, high altitudes, and northern Midwest, to nearly tropical in the latitudes of the deep South—the United States boasts an amazing diversity of landscape features, topography, soil types, and microclimates. With an impressive history of recent* glaciation, as well as many floodplains created by abundant water flows, soils are considerably more productive and diverse in the United States compared to Australia. Almost any sort of vegetation can find a niche in North America, so cultivation of anything even remotely suited to domestication, be it indigenous or of foreign origin, is readily possible somewhere in the United States. Moreover, this diversity of habitats provided what might have seemed like an unlimited food supply to those early pioneers. If we are to pinpoint one of the most important differences between Australia and the United States, it is the latter's tremendous biological productivity.

It was not until just under 5,000 years ago† that the great expansion of indigenous cultures in North America began to settle down and create more sedentary societies that could engineer the local surroundings beyond the extermination of local fauna. Although in many cases, such as the assorted and densely populated cultures that developed in what is now California, these patterns were driven by intrinsic productivity and not by agricultural proliferation. The prodigious productivity of the California coastal region—from its rich, upwelling-fed coastal waters to its diversity of deep-soil valleys—provided a veritable cornucopia of foodstuffs that indigenous people quickly learned to exploit. Nothing motivates like hunger, with sex perhaps being a close second.

Other regions in North and South America are more celebrated for the complex indigenous societies that developed following the invention of agriculture. The fertile floodplains of the Mississippi River gave rise to a developed society with a structured political framework around 1,000 years ago. However, the most successful pre-European agricultural societies flourished south of the current US-Mexican border. Native squashes were cultivated perhaps as long as 10,000 years ago, and maize was cultivated in Panama by 7,000 years ago, with

*Late Pleistocene.

†Recent evidence suggests that Australian Aborigines also rapidly expanded their population size about threefold around the same time, most likely arising from a combination of climate change and technological breakthroughs.

other species in the Amazon about the same time. Within the borders of the present-day United States itself, agriculture began in earnest about 4,500 years ago with the cultivation of many local grasses, along with imported crops from farther south. Much farther to the northeast, complicated indigenous societies regularly planted maize, beans, pumpkins, tobacco, and squash. Indeed, if it were not for the agricultural prowess of those early agrarians and their willingness to share that knowledge, the original English and Dutch Puritans would have most assuredly starved at Patuxet.[23] While hunting and gathering were by no means supplanted by the mostly seasonal supply of food energy from cultivation, it is the evolution of a partially agrarian lifestyle that allowed these people to flourish in a way that the first Australians never really could.

However, the climatic and landscape diversity did not ensure the persistence of these great pre-European societies. As eloquently described by Tim Flannery[24] and Jared Diamond,[25] most of these pre-Columbian societies failed to endure, collapsing utterly into ruin often mere centuries after their zenith. Even before *Homo sapiens* began to disrupt the climate, North America was also subject to extreme weather variation that brought droughts and floods to rival any environmental catastrophes observed in Australia, so the rapid rise and success of these early agricultural societies were generally followed by equally rapid declines and collapses. Local climate extremes probably contributed, but overhunting, deforestation, and unsustainable agricultural practices certainly played a large role, if not being the principal drivers of society failure.[26] While success built on the back of agriculture gave those early North Americans a decided advantage relative to Australian Aborigines, the latter's cultures endured for much longer. We are therefore of the opinion that the very conditions that prevented Aborigines from developing complex and densely populated (epidemic-prone) societies were also those that ultimately saved them from destruction by their own hand. That their unproductive landscape prevented early Australians from increasing their populations to the point of destroying their own environments, as so many other cultures have,[27] means that these limitations were actually their savior. Our modern societies would do well to limit their own economic, population, and consumption growth lest we duplicate the numerous collapses suffered by our ancestors.

Ironically, the high frequency of environmental extremes that contributed to the Aborigines' relatively slow pace of development was a phenomenon to which they must have adapted many tens of thousands of years ago. On the other hand, pre-European societies in America obviously benefitted enough from regional climate stability to develop complex agrarian societies quickly, only to plummet down the precipice of natural-resource collapse when conditions changed. So even thousands of years before European society was able to produce seafaring explorers, societal collapses in the New World were occurring with alarming frequency. With better technologies and foreign crops driving the recent success of Australia and America, are we merely delaying the inevitable crash? We argue in subsequent chapters that the modern halcyon period itself is at a close.

The next big wave in human colonization in the Americas and Australia began when Europeans decided that they needed a little more space and more resources to pilfer from others. But the motives of Europeans going to the Americas and Australia were very different. In the first case, it was essentially in search of riches; in the second, it was to create a sort of landfill for convicts. Ignoring a few early Norsemen attracted from Greenland to North America perhaps largely by the lure of trees and cod, Columbus and those who followed him were interested in profit and plunder. They were happy to take it at the expense of the local folks, and early successes provided Spain with enough silver and gold by the late sixteenth century to help maintain her as a major political power for a couple centuries to come.

The Europeans brought with them allies and technologies that utterly transformed the environment of the New World. The allies were a bunch of microbes—the agents behind diseases such as smallpox and measles—that promptly wiped out dense populations of Native Americans as effectively as those first people had wiped out the megafauna thousands of years before. About 90% of those first-contact Native Americans died in the European-introduced microbial onslaught. While we know a lot less about the epidemiological situation of Australia, we do know that Aborigines also succumbed to European epidemic diseases such as smallpox, whooping cough, measles, influenza, diarrhea, tuberculosis, and even the common cold. In 1789, a year after the First Fleet arrived, a smallpox outbreak killed many of the Aborigines who lived around Sydney, and then rapidly spread farther afield.

Overall, the unintended European biological warfare was probably almost as effective in Australia as in the Americas, albeit not as well documented.

European imports and technologies into the Americas included slaves, horses, cattle, weeds, saddles, firearms, steel-bladed plows, and railroads, all of which contributed to the total reworking of the ecology of the Western Hemisphere. Monocultures of cotton, tobacco, rice, sugarcane, and (largely in Central America) bananas generated fortunes and transformed landscapes, also often exhausting soils and restructuring economies and cultures. The role of the appalling, triangular slave trade in producing economic growth and prosperity for western European nations is, sadly, not often understood by the descendants of the slavers.[28] Slavery in Australia was on a much smaller scale—but perhaps 100,000 South Sea islanders were kidnapped and forced to work on sugar plantations, making a much smaller contribution to the prosperity of Australia, but lasting into the twentieth century.

The European invaders turned much of the grasslands of western North America into desert cattlescapes, simultaneously generating legends of the "Old West" that continue to be treasured by welfare ranchers today, some of whom are still engaged in promoting desertification in some areas while fortunately fending off suburbanization elsewhere. Unlike Australia, however, North America had a relatively well-watered interior, with large areas of deep, rich soils produced by extensive recent glaciation, as well as large tracts of forest. Early on, the interior's biological riches started to be mined. Up to the early 1800s, beavers—abundant wherever there was running water to be dammed and wood for damming—were the most valued commodity in the interior, trapped mercilessly for their hides, which were used in clothing. Millions of pelts moved in commerce, and beaver trapping* (a prime activity of so-called "mountain men") was a major factor in the exploration of North America. Beavers were wiped out over much of their former range, but happily with a declining market for furs and state-regulated trapping, they are recovering and reoccupying much of the suitable habitat that still exists.

*If the carnage had been restricted to beavers, the biodiversity toll would have been bad enough; however, many other fur-bearing species were targeted, such as pine marten, mink, and lynx.

What happened to the beavers was similar to the fate of a superabundant aquatic resource available in North American waters: the Atlantic cod. Cod is a fish that, unfortunately for the species, lends itself well to commercial exploitation. It is almost fat-free and preserves very well when salted and dried. Best of all, it tastes good dried despite having the consistency of a plank of wood. The Basques of northern Spain developed a thriving trade in cod a thousand years ago, and the fishes were harvested off Newfoundland as early as the late 1400s. Cod was once so abundant in the North Atlantic fishing grounds and so fecund that the author of a French cookbook wrote, with obvious hyperbole, that if all their eggs reached maturity, "it would take only three years to fill the sea so that you could walk across the Atlantic dryshod on the backs of cod."[29] But overfishing by commercial bottom trawlers caused a sudden crash of the population, reducing it to "commercial extinction" (too few animals to make it profitable to hunt them) in 1992. The number of harvestable fishes at that time was estimated to be 3 *billion* fewer than in the early 1960s, and the spawning biomass to be much less than 2% of the 1960s figure. Cod and other groundfish (commercially valuable species that live on the ocean bottom) populations now seem to be recovering, and we will see if a sustainable fishery can be established, at least for a while, in the face of climate change and ocean acidification.

In the interior, the iron horse transformed the North American landscape. With the help of the railroads, much of the agricultural activity shifted from the East Coast to the interior and the West Coast in the late nineteenth and early twentieth centuries. In the process, rich natural resources were overexploited. The railroads permitted billions of passenger pigeons to be slaughtered in the upper Midwest for food and as shooting-gallery targets, driving the species to extinction by the early twentieth century. The ecocide did not end with the pigeons—millions of bison were killed by market hunters, bringing that species to the brink of extinction. This was done with the direct approval of the US government, largely in an attempt to weaken plains native people, who depended on the bison for food and hides (and who harvested hundreds of thousands per year themselves). These slaughters, the plowing of the plains, and the grazing of massive herds of cattle (ecologically distinct and not as well adapted to the plains as the less water-dependent bison) continued the anthropogenic ecosystem re-

The restored mining ghost town of Bodie, California, near the Nevada border in the eastern Sierra Nevada mountains. The western push by the pioneers drawn by the gold rush laid ruin to much of the arid lands east of mountains. Photo by Corey J. A. Bradshaw.

organization that started with the overkill of the megafauna thousands of years earlier.

The end result was a rich continent more intensely altered by the activities of European immigrants than the poor continent of Australia. When flying over the inland areas of the United States, one cannot help but be impressed (or depressed, depending on your point of view) by the signs of human activity everywhere. No huge stretches of trackless "bush" are ever in view. Even the extensive forests of the Rocky Mountain region are crisscrossed by roads and dotted with cleared areas containing towns, subdivisions, ranches, ski areas, mines, and vacation homes. The center of the continent is a vast sea of crops, divided into giant squares by straight roads bordering "sections" (1-by-1-mile, or 258-hectare, plots), and punctuated by numerous towns. In the high plains, there are great circular patches of agriculture, the imprint of farming enterprises living on circular irrigation systems, pumping dry the greatest aquifer in the United States, the Ogallala. The rail lines that aided in the destruction of the passenger pigeons moved west in time to assist the bison slaughter and eventually connected the coasts in 1869. They proliferated into a vast web covering every state—today there is an estimated 228,000 kilometers (142,000 miles) of working rail lines in the United States.[30] In striking contrast, Australia did not finish an east-west transcontinental line until 1915 and a north-south line until 2007, still essentially the only rail lines outside of the eastern third of the Australian continent. By the late 1990s, Australia had only around 9,000 kilometers (5,600 miles) of working rail line,[31] and this network has been declining ever since. So relatively unimportant is rail transport in Australia that C.J.A.B. recently saw a sign on the back of a transport truck proclaiming: "Without trucks, Australia stops." It is sad claim, but essentially true.

Perhaps the greatest contrast with Australia is provided by North America's urbanization patterns. Over half of the Australian population now lives in five great cities, and almost all the rest reside in urban areas near them, all essentially coastal (see map on inside front cover). There are no Aussie equivalents of inland metropolises like the following US cities:

Chicago, Illinois (2.8 million people)
Phoenix, Arizona (1.6 million)
San Antonio, Texas (1.4 million)
Dallas, Texas (1.3 million)
Detroit, Michigan (0.9 million)
Austin, Texas (0.8 million)
Columbus, Ohio (0.8 million)
Indianapolis, Indiana (0.8 million)
Fort Worth, Texas (0.7 million)
Memphis, Tennessee (0.7 million)
Denver, Colorado (0.6 million)
El Paso, Texas (0.6 million)
Las Vegas, Nevada (0.6 million)
Louisville, Kentucky (0.6 million)
Milwaukee, Wisconsin (0.6 million)
Nashville, Tennessee (0.6 million)
Oklahoma City, Oklahoma (0.6 million)
Atlanta, Georgia (0.5 million)

Railway networks in the United States (*top*) and Australia (*bottom*). Note that the connection between Adelaide and Darwin was completed in 2007.

Albuquerque, New Mexico (0.5 million)
Kansas City, Missouri (0.5 million)
Mesa, Arizona (0.5 million)
Omaha, Nebraska (0.5 million)
Tucson, Arizona (0.5 million)

All those inland American cities (see map on inside back cover) are much larger than the capital of Australia, Canberra (0.4 million), which is basically coastal; and they are *gigantic* compared with Australia's only really inland town, Alice Springs (30,000). Even the above numbers for US cities misrepresent the original popularity of the central states, which have been steadily losing population to the South and periphery in recent years. In 2013 the city of St. Louis, close to the center of the continental United States, had 318,000 people,[32] down from 856,000 in 1950.

All of this means that while both Australia and North America were transformed environmentally by early human hunters exterminating many large animals, followed by agrarian overgrazing, only inland North America has been substantially plowed, replumbed, and built-over for agriculture and industry. In contrast, Australia retains a largely "empty" center,* dictated by the physical and biological factors discussed previously. Of course, this description ignores the huge areas of western Queensland, northern South Australia, and inland Victoria and New South Wales that are highly modified by sheep and cattle grazing. The effects of overstocking also occurred in Australia, but at an even higher environmental cost than in the United States. North America has long been an evolutionary crucible for hard-hoofed (ungulate) grazers and browsers. Many have gone the way of the mammoth, but surviving species include bison, deer (in all their varieties—from tiny key deer, a race of white-tailed deer of the Florida Keys, to giant moose farther north), antelope, and various mountain goats and sheep. In other words, North American ecosystems evolved with these hard-footed vegetarians—Australian ecosystems, on the other hand, did not.

Australia is famous for its "strange" fauna, none of which has hard feet. The grazers and browsers include various kangaroo species (including those that live in trees, rabbit-size wallabies, and powerful "nail-tails"), wombats, emus, and possums. Even most of the now-extinct megafauna were relatively light on their padded marsupial feet. Australian ecosystems thus evolved without this kind of pressure on the ground, and so the continent's delicate and nutrient-poor soils were overwhelmed once the hard-clod foreign beasts arrived. But it was not just the millions of cattle and sheep that quickly spread across Australia (today there are approximately 20 million head of cattle and

*The adjective "empty" does not really do justice to the tens of thousands of mainly Aboriginal Australians still residing in the deserts of central Australia.

A feral piglet in the Northern Territory. Do not be pacified by his cuteness—this is a massively destructive species in Australia. Photo by Corey J. A. Bradshaw.

about 70 million sheep roaming the country's rangelands); to Australia's ecological embarrassment and a monument to its lust for land transformation, today there are about 14 million feral pigs, over 1.5 million dromedary camels, 350,000 Asian swamp buffalo, 2.5 million feral goats, and 200,000 different deer (from six species) roaming wild across most of the country.[33] Pigs have ripped apart huge swathes of Australian bushland, given their propensity to root for subterranean delicacies (yams, freshwater turtles, etc.), and camels are munching on rare desert plants and dispersing many weeds in their droppings. Swamp buffalo have destroyed sensitive wetlands in the tropical north, and feral goats have knocked the fragile arid lands into a state of potentially irreversible degradation.

The combination of ecological ignorance and disrespect by the Europeans was not restricted to nonhuman animals and plants—European invaders of both continents had a record of execrable treatment of Aboriginal and native populations, often simply attempting (with some success) to exterminate those who survived the assault of the foreigners' microbial allies. In North America, relatively sustainable hunter-

Feral swamp buffalo *Bubalus bubalis* originally from Southeast Asia cause immense damage to billabongs (waterholes) in northern Australia. Photo by Corey J. A. Bradshaw.

gatherer and subsistence-agriculture societies were destroyed. They were replaced over large areas with increasingly industrialized monocultures of plants and animals designed to supply urban centers and international trade. On neither continent was there any real attempt to integrate indigenous people into either the agrarian or urban economic sectors. In the United States, much of the farm labor through the mid-nineteenth century was supplied by Africa-derived slaves, and in both countries a steady stream of European (and in the United States, Asian) immigrants helped fill the need for cheap labor for farming, factory work, and railroad construction. While many Aborigine jackaroos* became essential to the expansion of livestock on outback stations (ranches) in the early part of the twentieth century, most were summarily sacked following equal-pay laws that were passed in 1966 (the official excuse was that wealthy white farmers could no longer afford the once-cheap labor—the deleterious societal effects of that

*Stockman; "jackaroo" is a term apparently derived from an Aboriginal language, ironically meaning "wandering white man."

mass firing are still rife today). Exploitation of the poor was matched throughout by exploitation of the land, culminating in North America with the "dust bowl" of the 1930s, which destroyed the productivity of millions of hectares of farmland. Better farming techniques were introduced, but soil erosion, over-pumping of aquifers, and climate disruption all continue to threaten the artificial ecosystems that occupy most of the center of the North American continent. The recent drought experienced by the central and western United States in 2012 might herald a second dust bowl, eventually leading to another wave of food shortages.[34]

What is remarkable about the neo-feudal land-grabbing of the American frontier is how long it lasted—nearly 300 years of acquisition (often accompanied by military force), transformation, stocking, and irrigation. Australia was colonized much later than the United States (only about 200 years ago), and so most of the land transformation occurred later (and faster) in the late nineteenth and early twentieth centuries. But again, the history of land monopolization was strikingly different—Australia's first colonists were by and large ex-convicts who, following Britain's eventual disinterest in the clerical and financial management of its distant penal colonies, were allocated parcels of land to farm. Those of the more successful early ex-convicts, and many later arrivals, were drolly dubbed the "squatocracy" because of their eventual rise to economic power following the acquisition of essentially free land. The only exception to this general trend was in the state of South Australia and its main city, Adelaide, where C.J.A.B. lives; most Adelaideans are still today proud to affirm their non-convict past. In that state, mainly English and German agricultural entrepreneurs quickly established successful wheat and wool markets to the almost-complete demise of its forested ecosystems. As a result, South Australia is today one of the most ecologically devastated states,[35] despite its relatively small human population of 1.7 million[36] (as of December 2013).

While the United States eventually grew to be a world power based on the foundational exploitation of its abundant natural resources, Australia did not develop as a major trade center throughout most of its European history. Of course, it successfully exported sheep and grain (mainly to the United Kingdom), but the country was never under pressure to become a big player in manufacturing. Even today Aus-

Fight between Aborigines and Mounted Whites by Samuel Calvert (1828–1913). Reproduction rights courtesy of the Mitchell Library, State Library of New South Wales.

tralia is still very much a resource-based economy, but the emphasis has shifted from agricultural products to mining exports—coal, uranium, iron ore, bauxite, nickel, zinc, and natural gas now keep her citizens fat* and wealthy. Ironically, this has made Australia do its best to emulate a developing nation entirely dependent on exporting its raw resource wealth. With the exception of coal, which spreads its influence across the entire planet by disrupting our climate, the ecological footprint of mining is probably less than that arising from the deforestation required for broad-scale agriculture. However, the ecological devastation of agriculture has already happened in both Australia and the United States—damage that is still largely unrecognized by their human populations.

*Australia is ranked as one of the fattest nations in the developed world, with over 60% of all Australians considered overweight or obese (source: www.modi.monash.edu.au/obesity-facts-figures/obesity-in-australia). In the United States, over 67% of adults and 33% of children aged 6–19 years are overweight or obese (source: win.niddk.nih.gov/statistics).

3 Remorse

One of the penalties of an ecological education is that one lives alone in a world of wounds. Much of the damage inflicted on land is quite invisible to laymen. An ecologist must either harden his shell and make believe that the consequences of science are none of his business, or he must be the doctor who sees the marks of death in a community that believes itself well and does not want to be told otherwise.

ALDO LEOPOLD, *A Sand County Almanac* (1959)[1]

The very different human histories and topographies of Australia and the United States have naturally led to rather different environmental histories and approaches to related problems. When the two nations emerged from the heyday of the exploitation party, they awoke with colossal environmental hangovers and the inevitable remorse that followed. Europeans invading Australia had found a continent that had been occupied for some 50,000 years by hunter-gatherers and a sparsely occupied desert inland. Europeans invading North America found large, native agricultural communities and a seemingly endless rich hinterland. With the Native Americans conveniently and nearly annihilated by diseases that accompanied the invaders, the Europeans made relatively short work of the survivors and then went on to exploit the natural riches. This provided immediate lessons in the finite nature of even an apparently unlimited natural bounty. In a few decades of the railroads penetrating the US interior, the passenger pigeon had been forced to extinction by market hunters, with the bison nearly following them to their evolutionary grave. In contrast, exploration of the outback of Australia revealed no obvious abundant resources and no hoped-for inland sea. There was no natural highway like the Mississippi River to contribute to the commercial exploitation of the Red Centre. Australia therefore had a more difficult job of "developing" according to the romantic ideals of its European colonists—this also contributed to the late birth of environmentalism there.

It might be a surprise to even the most "green" of you reading this book that an important stage in the development of environmental consciousness in the United States was based on the wanton exploitation of an unusual resource—the feathers of birds. Early conservationist Frank Chapman famously "birded" in Manhattan in 1886 and twitched* forty species of native birds—all identified in the feathers adorning ladies' hats. The feathered-hat craze of the late 1800s and early twentieth century was causing a boom for plume† hunters, who raided wader and gull rookeries along the East Coast of the United States. They nearly exterminated the great and snowy egrets, which were especially prized for their soft, long, breeding plumage. The avian massacre, with millions being slaughtered annually, gradually built outrage in the American population. In response, societies were formed to work to protect the birds, and laws were passed to regulate hunters and the feather trade. Those societies took the name of the great ornithologist/bird artist, John James Audubon. Thus was born the great Audubon Society, which is still a force in North American nature conservation.

The landmark piece of US federal conservation legislation was the Lacey Act[2] passed in 1900, which prohibited interstate transfer of wild animals killed in violation of state laws. Pressure from the Audubon societies then forced state after state to pass equivalent laws. Gradually the feather craze faded away, in part because of the changing legal landscape, in part perhaps because of rising prices, changing fashion trends, and modernizing roles of women. By World War I, displaying feathers was once again just an avian trait, but the episode marked the founding effort in conservation in the eastern United States. In the west of the country at about the same time, the wandering naturalist, philosopher, and author John Muir was building a national consciousness of the glories of the Sierra Nevada mountains and similar natural wonders. His activities led directly to the establishment of California's

*For those uninitiated into the strange world of crazy bird-watchers (like P.R.E.), a "twitched" bird is one observed and recorded as such by the "twitcher" (bird observer).

†Plume (French) = feather.

Yosemite National Park in 1890 and several other western parks subsequently, which arguably gives him the unofficial title of "father" of the US national park system (although Yellowstone National Park was established first in 1872). Perhaps even more important was his role in the founding of the Sierra Club in 1892 and becoming its first president. Gradually at first and then rapidly thereafter, it became a powerful force for conservation; like the Audubon Society, it remains so today.

This was also the era of America's first conservation-oriented president, Republican Theodore Roosevelt. A big-game hunter, he had become distressed at the overgrazing and other environmental degradation he had observed in the West. When he became president in 1901, Roosevelt wielded his considerable power to protect wildlife and public lands. He established the US Forest Service and set up 51 federal bird reservations, 4 national game preserves, 150 national forests, 5 national parks, and through the 1906 American Antiquities Act created 18 national monuments. Overall, he protected more than 80 million hectares (200 million acres) of public lands, and established a connection of the Republican Party with environmental protection that was enhanced later by none other than Richard Nixon, but then the trend was reversed by Ronald Reagan.

While all of this was going on, there was a little-remembered, but large-scale progressive movement to popularize a conservation ethic. This "nature study movement"[3] promoted outdoor education for youths—including field trips, building and emplacing birdhouses, gardening, and photography—to help connect them with rural values and the natural world and increase their joy of living. Would that both the United States and Australia had something similar still today! Well-known scientists involved included Louis Agassiz, who was a fine naturalist but sadly a holdout against Darwinism, and Anna Botsford Comstock, who did wonderful wood engravings of insects for books by her husband, John Henry Comstock. Other prominent figures involved included Booker T. Washington and Ernest Thompson Seton, whose book *Lives of the Hunted*[4] introduced P.R.E. to conservation ideas as a teenager. Perhaps P.R.E.'s best memory of these pioneer conservationists is of reading naturalist/photographer Gene Stratton-Porter's *Moths of the Limberlost.*[5] Her joy in the beauty of the moths was balanced by careful explanations of how her own observations often

Vista of Half Dome in Yosemite National Park in the Sierra Nevada mountains of California. Photo by Corey J. A. Bradshaw.

countered statements in standard books about moths. When gorgeous large saturniid (silk) moths emerged during the day from cocoons P.R.E. had collected in the woods, his mother would drive to his junior high school to pick him up so he could get home early and enjoy watching their beauty unfold.

It was also in the United States that the issue of ubiquitous toxins was first introduced to the general public in 1933. Arthur Kallet and F. J. Schlink wrote a book entitled *100,000,000 Guinea Pigs: Dangers in Everyday Foods, Drugs, and Cosmetics.*[6] It was far ahead of its time, discussing such topics as the impacts of long-term exposure to low doses of toxins such as pesticide residues and food additives, synergisms among the many toxic substances to which people are exposed, the sleazy behavior of corporations, and corporations' regulatory capture of the US Food and Drug Administration. Listen to their voices from the past:

> Using feeble and ineffective pure food and drug laws, the food and drug industries have been systematically bombarding us with

> falsehoods about the purity, healthfulness, and safety of their products, while they have been making profits by experimenting on us with poisons, irritants, harmful chemical preservatives, and dangerous drugs.

Throw in the tobacco and chemical industries, and these lines would acquire a twenty-first-century aura. Or consider their statement:

> We see the law pay little attention to the civil rights of the small businessman and of the individual as consumer, at the same time that Federal, State, and city officials touch corporation matters with extreme delicacy, restraint, and secrecy.

That could be a comment on Antonin Scalia and his fellow judicial thugs on the *Citizens' United* Supreme Court case,[7] or then–vice president Cheney's secret 2001 conferences with energy barons.[8] Kallett and Schlink's eighty-year-old summary sentences also could almost apply to the toxics situation today:

> Even in *laissez-faire* America, we have found that we simply cannot permit ignoramuses to build and operate steamships or to build railway bridges. To permit the uncontrolled manufacture and distribution of foods and drugs is a far more reckless disregard of the people's rights.

When P.R.E. was a graduate student in the Department of Entomology at the University of Kansas, the pesticide DDT* was very much on his mind. His assistantship in 1953 involved doing research on the evolution of DDT resistance in fruit flies, under the direction of the late, great Robert Sokal, the father of numerical taxonomy and the scientist who brought statistics to biologists. Under Bob's tutelage, it quickly became clear to all in his research group that the broadcast use of pesticides was a losing and dangerous game. P.R.E. was primed to accept it; when he had attempted to raise butterflies in New Jersey in the late 1940s, bringing food plants in from nature usually resulted in the caterpillars dying. In those days, widespread spraying of DDT to con-

*Dichlorodiphenyltrichloroethane.

trol mosquitoes coated much of the countryside with poison. In the lab, P.R.E. found it easy to use selection to make flies impervious to DDT in some ten generations, or, in contrast, so susceptible that they would drop dead at a whisper of that "miracle" chemical's name. Evolution of resistance makes continuous use of any pesticide near certain to reduce its efficacy. The usual responses of the chemical industry, recommending increasing the dose or substituting even more toxic compounds, tends to make control both more expensive and more dangerous. It is a recipe for disaster in the long run.

That was well understood by evolutionists early on, but it took a marine biologist and talented writer, Rachel Carson, to bring the pesticide problem to public attention and, incidentally, to launch the modern environmental movement. *Silent Spring*[9] is a brilliant book, a devastating indictment of the then-common (and still today—see chapter 6) misuse of pesticides, but also one that appeared when the time clearly was ripe. The public seemingly had been primed by publicity about radioactive fallout, fears of pesticide residues on cranberries, and the thalidomide drug scandal,[10] the latter enhanced by pictures of infants born with distorted limbs after their pregnant mothers had taken thalidomide to prevent morning sickness. Carson suffered from the drawbacks of being a female scientist before science's gender gap began to narrow and from lacking a PhD and a professorial position. Unfortunately, that gap still has quite a long way to go to closure even today. Despite those disadvantages, she had the science about as right as it could be at the time. She discussed the developing problem of the evolution of resistance to pesticides, something still not adequately dealt with by producers or users, or by the medical establishment, which had been long facing parallel problems with antibiotic resistance. She understood the problem of pesticide use in promoting innocuous organisms to pest status, and the concentration of pesticide residues in food chains. She also described what was then known about the hazards that biocides represented to both wildlife and human health.

Carson was subject to a storm of vicious attacks by the combined public-relations machines of the chemical industry and agribusiness, and even from scientists in the industry and in entomology departments of some universities. Typical was the much-quoted statement of Professor Robert H. White-Stevens, a "poultry scientist" at Rutgers University:

Portrait of Rachel Carson by Minnette D. Bickel (1987). Article Photograph of Oil Painting. Series: Pennsylvania College for Women, Chatham College, USA. Permission to reproduce courtesy of the Jennie King Mellon Library, Chatham University.

> If man were to follow the teachings of Miss Carson, we would return to the Dark Ages, and the insects and diseases and vermin would once again inherit the earth.

The assault began even before *Silent Spring* was published, and the baseless defense of planetary poisoning by chemical-industry representatives continues today.

Velsicol Chemical Corporation, a manufacturer of chlorinated hydrocarbon pesticides, once threatened to sue P.R.E. for what he had written about the company's behavior and the honesty of its executives. P.R.E. felt honored since Velsicol had threatened a libel suit against the publisher Houghton Mifflin before *Silent Spring* saw print. A public-relations campaign against Carson and trumpeting the safety of pesticides was bankrolled by the National Agricultural Chemicals Association. Virtually all of the attacks were without merit. Carson withstood them all, even though she was struggling with metastatic cancer. P.R.E. met her once only briefly, and then she was gone.

But her legacy looms huge today. Many people have the impression that climate disruption is the worst environmental problem humanity faces, and, indeed, its consequences will likely be catastrophic. But of other candidates for the biggest threat, the spread of toxic chemicals from pole to pole could be the dark horse in the race. Carson might have started environmentalism by illuminating exactly the right issue, at exactly the right time.

The part of US civil society dealing with environmental issues had expanded following Carson, in part promoted by the upheaval caused by the United States' involvement in the Vietnam War.* Plans were made for the first Earth Day on April 22, 1970. For the most part, it was a plea for more government action to protect the environment. Another Republican president, Richard Nixon, saw the trend, and although he is now mostly remembered for his Watergate† crimes, his administration was the first since Teddy Roosevelt to carry forward a series of pro-environment initiatives. Three months before Earth Day, in his first State of the Union speech, he said:

> Restoring nature to its natural state is a cause beyond party and beyond factions. It has become a common cause of all the people of this country. It is a cause of particular concern to young Americans, because they more than we will reap the grim consequences of our failure to act on programs which are needed now if we are to prevent disaster later.

In 1970 his initiatives included establishing the Environmental Protection Agency and supporting the passage of the Clean Air Act. He also moved to establish new national parks and during his presidency launched many other environmental proposals. In 1973 he signed the Endangered Species Act, perhaps the most significant of the many en-

*Which the Vietnamese to this day call the "American War."

†In 1974 Nixon resigned because of his involvement with a 1972 break-in into Democratic National Committee headquarters

vironmental laws passed in the United States in the 1970s. Another courageous action by Nixon, with highly important environmental repercussions today, was regularizing relations with the People's Republic of China. We say "courageous" because he had to counter a powerful right-wing lobby backing the dictator Chiang Kai-shek, who had retreated to Taiwan. All in all, it was a pity that Nixon's less admirable characteristics led to him leaving office in disgrace.

The Nixon era put the United States into a position of world leadership in the environmental movement, which started to become international with the founding in Canada of the aggressive organization Greenpeace in 1971. The Vietnam War created nearly the same turmoil in Australia that it did in the United States, although civil society there had not been energized by the prospect of nuclear fallout and Carson's plea to end the tyranny of toxification. In Australia there were no equivalents of the Sierra Club and Audubon Society. In fact, Australia's awakening to environmentalism occurred for much different reasons (see later in this chapter).

More recently, the United States has fallen far behind on the environmentalism front, despite the efforts of a wide variety of environmental nongovernment organizations such as the Sierra Club, World Wildlife Fund, Environmental Defense Fund, Natural Resources Defense Council, Audubon Society, Union of Concerned Scientists, Nature Conservancy, and many others. That was largely due to the election of the person likely to prove America's most dangerous president ever, Ronald Reagan; with his taking office in 1981, the Republican Party parted definitively from its past environmental agenda and credentials. One of his first acts in office was to shred every copy his minions could lay their hands on of the *Global 2000 Report to the President.*[11] The report was a landmark effort commissioned by President Jimmy Carter as a cooperative venture by US government agencies to look toward the environmental future. Reagan also removed from the roof of the White House the solar panels that Carter had installed there; they have only just been replaced by the Obama administration. Reagan was probably suffering from dementia when he was president, but whatever his excuse, he managed to initiate a decline in Republican support for environmental action in the United States at a critical point, a decline practically complete when Obama was elected. Reagan thus contributed greatly to the disintegration of the nation's (and the world's) natural life-support systems. His attitude toward the environment was well-encapsulated in his infamous public statements: "trees cause more pollution than automobiles do" and that if "you've seen one tree you've seen them all." His acts matched his words.[12] As a representative of the Wilderness Society put it:

> The Reagan administration adopted an extraordinarily aggressive policy of issuing leases for oil, gas, and coal development on tens of millions of acres of national lands—more than any other administration in history.

Reagan selected two anti-environmental operatives, James Watt and Anne Gorsuch Burford, to head the Department of Interior and the US Environmental Protection Agency, both of whom spent their tenures in office trying to roll back environmental protections. He and they set the tone for a subsequent Republican war on the environment that continues to this day, disgracing the party of Theodore Roosevelt.

The Australian groggy awakening to its environmental dilemmas was very different from that of the United States. Not only did Europeans arrive in Australia en masse only about 200 years ago—much later than their American counterparts—they had a more limited view of their new landscape and were, at least initially, constrained by the harshness of their new home. Those mostly British settlers brought with them the fully formed ideas of development and progress shaped by centuries of land use in the motherland. That ideal of conquering wilderness and transforming it into the bucolic landscape typical of the English countryside was their driving force. The early settlers viewed the Australian bush as ugly and monotonous, features that could only be overcome by human occupation and cultivation.[13] This neoclassical view, homesickness, and the Romantic desire to transform their homes and farms into an image of those from their homeland were defining forces in early Australian history. Unlike in Europe, though, where there were cultural taboos associated with forest degradation—bound in mysticism, spirituality, folklore, and politics—no such restrictions applied to the unfamiliar Australian bush.[14] In fact, the Australian government passed the Crown Lands Alienation Act in 1861, which was designed to "open up" the colony to settlement and penalized landholders for *not* clearing the land (via a forfeit of the

Clearing for Agriculture in Early Settlement by anonymous, Government Farm at Castle Hill (ca. 1803), watercolor, 24 × 35 cm. From C. Jordan, "Progress Versus the Picturesque: White Women and the Aesthetics of Environmentalism in Colonial Australia, 1820–1860," *Art History* 25 (2002): 341–57. Permission to reproduce courtesy of the Mitchell Library, State Library of New South Wales.

land back to the Crown).[15] That single act guaranteed the deforestation wave would continue for over a hundred years.[16] That, and the persistent desire to make the new land look as much as possible as the old homelands has ensured that continuing demise of Australia's biowealth.

Interestingly, and in many ways comparable to the United States' war on Native Americans for their land, clashes over land use between the settlers and Aborigines were probably some of the first demonstrations of what today we would call "environmentalism" in Australia.[17] Aborigines were intent on preserving their way of life (and, indeed, their *lives*) in the face of the settlers' onslaught. But this was seen, at most, as a mild inconvenience for the new Australians, who in response invoked the idea of *Terra Nullius*—that no one owned the land, making it available to anyone (white) who wished to "improve" (clear) it.[18] As we discussed in previous chapters, Aboriginal Australians, like their Native American brothers and sisters, decidedly lost that battle.

In fact, it was not until 1972 that the landmark High Court ruling in the famous *Mabo vs. Queensland* case gave native title a real legal footing in challenges to public land ownership.[19] While indigenous organizations today directly or indirectly own about 20% of Australia's land,* claims by Aboriginal groups invoking native title laws are still a common occurrence today.

But protectionist notions were not restricted to the Aboriginal struggle for rights and recognition at this time; there were in fact smatterings of utilitarian environmentalism expressed even among the early European governors. Even as far back as 1803, and only five years after the First Fleet landed in the area that was eventually to be called Sydney, the Governor of New South Wales, Philip Gidley King, ordered that trees along the Hawkesbury River were not to be cut down due to erosion concerns. His orders went largely unheeded by the development-hungry colonists.[20] Later in 1865, a report issued to the Victorian Parliament complained that overexploitation of forests would soon leave little timber for the requirements of other industries such as mining. Other government botanists, clergyman, doctors, private citizens, and even early feminist artists[21] also complained bitterly about the wanton destruction of forests as the colonial wave passed through the country.[22] For the most part though, their cries went unheard or were completely ignored.

Australia's first national park was created in 1879 on the outskirts of Sydney. While the establishment of the 7,300-hectare (18,000-acre) "National Park" (as it was known, now the Royal National Park) was most likely inspired by the creation of Yellowstone National Park in the United States seven years earlier, the two areas had very different purposes. Yellowstone was created primarily to preserve its natural wonders from development for the benefit of all Americans, whereas the National Park in Australia was created primarily for the sport and recreation of the 200,000 citizens of the colony's largest urban center, Sydney, only 20 kilometers (12 miles) to the north. As such, preservation was an extremely low priority, with uses such as ornamental plantation, zoological gardens, race courses, artillery ranges, mining, timber felling, livestock grazing, and deer husbandry all permitted.[23]

Although these decidedly non-conservationist activities were even-

*In the Northern Territory, it is closer to 50%.

tually excluded, their legacy still weighs heavily on the region's biowealth. Subsequent parks were established around the country and, as with the National Park, primarily as recreational respite from urban living around the country's major population centers—these included Belair in South Australia (1891), Ku-ring-gai Chase north of Sydney (1894), Witches Falls and Bunya Mountains near Brisbane (1908), Cunninghams Gap also near Brisbane (1909), Lamington near the Queensland–New South Wales border (1915),[24] and Mount Field in Tasmania (1915—later revoked). The first Australian national park created more in the spirit of Yellowstone was probably Mount Warning in the far northeast of New South Wales in 1929. This was followed by Dorrigo Mountain in 1930 and New England National Park in New South Wales in 1937. Australia's largest national park, Kakadu National Park in the far north of the Northern Territory, was declared in three stages between 1979 and 1991.

Many modern Australians have their own romantic notions of how those early European Aussies managed to break that English ideal and become the "true blue," bushranger types lauded in poetry, literature, and film. Even today, Australians have a comical love/hate relationship with the English—while still by and large accepting the Queen as the official (if essentially powerless) head of state, Australians are quick to label any English visitor the jocular soubriquet of "pommie" or "pom,"* which is only *mildly* derogatory. After shedding the perceived uppity and delicate sensibilities of the English gentry, the archetypal image of the hard Aussie "bloke" emerged. By the mid-nineteenth century, the view that Australia was a vast, untapped resource of mineral and agricultural wealth, ripe for exploration, gave the (mostly male) pioneers the aura of the imperturbable "frontiersman." Whether a miner, prospector, surveyor, trapper, or timber cutter, the noble pioneer fighting the harsh land to feed his family and do his bit for the development of the nation soon become the stuff of legend. Able to survive off the

*While the origin of this term is commonly thought to derive from the acronym of "prisoner of Mother England" (POME) or "prisoner of Her Majesty" (POHM or POM), it is probably more likely a derivation of "pomegranate," which referred to either an English settler's flush complexion after a bout of Australian sun, or as rhyming slang for "immigrant."

land with nothing more than a rifle, billy,* and swag,† the Aussie frontiersman is still celebrated in the unofficial national anthem and bush ballad—"Waltzing Matilda." In fact, the archetype persists even in today's eating habits—Australia is the only nation that regularly partakes in the consumption of its national emblems.‡ This image also contributed to the rise of unions in Australia the following century and is probably at least partially responsible for the now-dwindling socialist components of Australian governance today, including public health care and unemployment and housing subsidies.

The Second World War—and, to a lesser extent, the First World War—saw another change in the Australian culture for development. Following the economic lull of those periods of world crisis, Australia, like many other countries, adopted a hunger for land development that gave rise to a massive economic boom in the 1950s and 1960s. Aided by schemes to reward the returning Diggers with entrepreneurial and employment opportunities, development clicked up several gears, where the very notion of "natural resource" became synonymous with "commodity."[25] It is this era that spawned the massive vegetation-clearing schemes of Western Australia and Queensland, and the likes of corrupt politicians such as Queensland Premier Joh Bjelke-Petersen, hell-bent on removing as many trees as possible in the name of "progress." The only reason the Great Barrier Reef was spared Queensland's lust for exploration and mineral extraction was the intervention of the federal Labor government;[26] it was not because of popular protest that the Reef exists at all today. But even that success is under massive threat—nearly half of the coral in the Great Barrier Reef has been destroyed in the last thirty years.[27]

But it was the rising environmental collateral damage of this development push that finally persuaded some Australians to adopt the first sentiments of environmentalism. It might shock people to learn that Australian environmentalism's birth came so late, despite its early gestation in some societies and reserves, and having access to the literature and ideas emanating from the United States. The defining mo-

*Camping pot.

†Camping bedroll.

‡A kangaroo and emu face each other on the Australian coat of arms.

ment came about when a Tasmanian hydroelectric dam proposal threatened to flood a small lake—Lake Pedder—in the central highlands. The lake itself had been declared a national park in the 1950s, and so its threatened existence became a rallying point mainly for bushwalkers and field naturalists. Bushwalker clubs had become increasingly popular since the 1920s and formed a mounting voice in the preservation of natural landscapes.[28] But the outcry grew rapidly beyond that small demographic, with the lake becoming in 1967 the focus for a national debate that generated two state elections, a decade of state and federal lobbying, federal interference in Tasmanian state politics, and years of political vitriol.[29] Although the struggle to save the lake ultimately failed (it was flooded in 1972), ironically the affair ignited a wholesale paradigm shift in Australian land-use policy. Combined with the publication of the book *The Fight for the Forests*[30] by two Australian National University philosophers in 1973, these events sparked the creation of the world's first "green" political party—the United Tasmania Group, which contested the 1972 Tasmanian state election and dealt a major blow to the forest industry.[31] This is probably the one event in the history of environmentalism where Australia leapfrogged the United States. The United Tasmania Group also became the forerunner of one of the country's most prominent environmental nongovernment organizations, the Australian Wilderness Society, and the political party the Australian Greens.

Although national parks had a rocky start in Australia and were not created expressly for the purposes of wildlife conservation, they did provide the framework for eventual conservation legislation. The Wildlife Preservation Society was formed in 1909 to preserve intact the "typical fauna of Australia" in response to concerns about dwindling wildlife populations around Sydney. The next major form of legislation appeared in the New South Wales Fauna Protection Act of 1948, which governed land reserves around the state for protection and study of native fauna and flora, although it was not until 1967 under the auspices of the National Parks and Wildlife Service that fauna protection had any legal teeth at the federal level of government.[32] Finally, the Environmental Protection and Biodiversity Conservation Act 1999 (EPBC) came into existence as the major national legislation, which today enforces legal issues surrounding biodiversity conservation and restoration—it is the Australian equivalent of the US Endangered Species Act.

Although there was not much planning involved initially, Australia (like many other countries) started to take biodiversity conservation seriously in the mid-1990s, such that now there is about 11% of Australia's land area within a National Reserve System. Science-based planning to ensure that the system protected enough of the things most in need of protection did not feature heavily in the early years, such that today only 19% of the 2.93 million square kilometers (1.13 million square miles) of land considered "wilderness"* is protected in some form.[33] Likewise, Australia recently (2012) legislated the establishment of the world's largest network of marine reserves, but many environmental groups and scientists believe that it still falls well short of protecting the most threatened and unique areas of marine biowealth in the country. The bottom line is that Australia's reserves do not really capture what needs protecting, with recent protection given mainly to areas previously modified heavily by humans.[34] Even Australia's largest and best-known national parks are struggling to save the biowealth that they were created to protect.

As in many wealthy nations, Australia has no shortage of environmental organizations, from political parties such as the Greens, to national- and state-level nongovernment organizations like the Australian Wilderness Society, the Australian Conservation Foundation, BirdLife Australia, Greening Australia, National Parks Association, and to international organizations with Australian nodes such as the World Wildlife Fund, Greenpeace, and Conservation International. Australia also ratifies many international accords with respect to wildlife conservation, trade and environmental custodianship, including *inter alia* the Antarctic Treaty, the Kyoto Protocol, Convention on Wetlands of International Importance Especially as Waterfowl Habitat (Ramsar Wetlands), the Convention on International Trade of Endangered Species (CITES), the Convention on Biological Diversity (CBD), and the International Convention for the Regulation of Whaling. Of those treaties listed, only the Kyoto Protocol is not also ratified by the United States. In other words, Australia has a seemingly healthy

*Defined as "large areas that have experienced minimal habitat loss."

army of conservation-minded individuals, organizations, and political movements. So why are there still so many environmental problems, and why are they getting *worse* instead of better?[35]

While plutocratic politics and ideologies are certainly a major force, one reason there is an appalling lack of momentum on the environmental front in Australia is that mainstream environmental organizations still by and large occupy an antagonistic position with respect to major public policies. Being one of the world's most urbanized populations (nearly 90% of Australians live within a major urban center), there is still huge gulf between the "bush" and the "city"[36]—a notion that severely impedes the development of sustainable living, urban ecology, and a pan-Australian environmental movement.[37] While there have certainly been great leaps forward in the last decade, spurred mainly by the rising tide of concern regarding human climate disruption (e.g., the creation of a federal Department of Climate change in 2007, but its subsequent demise in 2013), Australia has not moved much beyond the "us" versus "them" (i.e., "dirty greenies" versus "capitalist pigs") paradigm that started the whole movement in the first place. This is unfortunately a rather juvenile approach to solving the world's tough environmental issues, and, if anything, the "us" versus "them" polarization and radicalization in environmental politics is getting worse with each passing year.

Countries like Australia and the United States are in dire need of a united front to tackle the massive anti-environment sentiment gripping many within their populations and the anti-science positions of major elements within their governments. The various environmental nongovernment organizations, state-centric bodies, lobby groups, and even political parties swing hither and yon when it comes to environmental policy. While there are sometimes temporary coalitions based on specific or local environmental issues, there is no charter of agreed environmental policy issues with which the many environmental organizations can use to lobby state and federal governments as a strong, united front. Despite this shortcoming, many major national groups have remained quite consistent and in agreement on important policies, although they may differ in their strategies and priorities. They also tend to carve up the territory; some litigate, some lobby, some use land purchases for habitat preservation; many focus on public education or grassroots campaigning.

The problem is, however, that despite the good work most of these organizations do on behalf of all of us for the protection of our life-support system, there are even more points of contention that must make those with any sort of "green" conscience look like a pack of squabbling camp dogs to the opposition. Never is the competition fiercer when the stakes are so low. But few realize that the stakes are not low—they are huge, and they will affect all of us, libertarian and communist, green and coal-brown alike. In other words, environmental organizations cannot afford to act like entrenched minor political parties if they want any real chance of taking environmental policy by the horns. We need a charter of agreed-upon principles and policies, with binding commitments not too unlike those of international agreements such as the Kyoto Protocol that each signatory agrees to promote and support. A massive league of like-minded organizations and individuals would make environmental issues mainstream ones and would avoid the rapid vortex of extinction-driving politics we are now facing.

In Australia in particular, there is more public concern for the plight of whales and the circus that is the International Whaling Commission, than the biodiversity tragedy unfolding in the nation's far north. While the iconic Great Barrier Reef is in the public eye, it takes attention away from the huge problem of illegal fishing in its northern Exclusive Economic Zone. While Australians over-allocate and fight for every remaining liter of water from the doomed Murray-Darling River (the nation's largest and most threatened freshwater system), there is a shocking silence on the destruction caused by feral cats to its marsupial fauna.[38] The indifference is alarming.

This indifference is at least partially a result of a clever political trick designed to distract the public from the major environmental problems, as the case of Australia's love affair with anti-whaling movement readily demonstrates. For years, Australia has been a self-righteous, loudmouthed condemner of whaling nations at the International Whaling Commission. While this considered in isolation is without doubt a laudable goal—of course humanity should not be hunting these magnificent marine creatures—it is one of the greatest environmental wool-pulling-over-the-eyes political sideshows ever devised.

By being so vocally anti-whaling, Australian politicians can win easy green votes while doing nothing much at all about the other, real

Aboriginal man from the Wessel Islands, Arnhem Land, in the Northern Territory of Australia delivering the final deathblow to a green sea turtle *Chelonia mydas* using a rock. The turtle had been caught days before via a spear to the shell and was cooked on an open fire and consumed that very evening. Photo by Corey J. A. Bradshaw.

environmental crises unfolding right beneath the noses of their constituents. It is an easy political gain—even the most hard-core, right-wing plutocrat would probably not, at least publicly, denigrate a government for standing up for the whales. Being anti-whaling is not a controversial environmental stance. With a little emboldened brinkmanship on the international stage, bolstered by some over-the-top, sensationalist media coverage, today's politicians have a guaranteed recipe to garner faux environmental kudos. It is a case of brilliant politicking and is entirely disingenuous. Australia was in fact a committed whaling nation up to 1979. Combine that historical nugget with Australia's legal kills of dugongs, sea turtles, and other native species under native sustenance laws, as well as state-sanctioned culls of predators like dingoes and sharks, and the anti-whaling bravado takes on an air of hypocrisy. Australia's terrible environmental record and systematic erosion of its hard-won environmental protection laws makes this stance laughable.

In the end, it is important to keep in mind that despite the struggles of many people in civil society and in government, the environments in both Australia and the United States, as in the rest of the world, have continued to deteriorate. Successes have been local and all too often temporary, while the disasters have tended to be global and permanent. It is crucial to keep in mind what the main global and national environmental issues are: loss of biodiversity, climate disruption, toxification, resource conflicts, and deterioration of the epidemiological environment, all primarily driven in Australia and the United States by increasing overpopulation and overconsumption.

4 Biowealth*

If we let a species go extinct, we have foreclosed on the possibility that we might discover the species to be important. We ought to preserve biodiversity to hedge our bets.

JAMES MACLAURIN AND KIM STERELNY,
What Is Biodiversity? (2008)[1]

Homo sapiens is a relatively new addition to the global pool of life that we call "biodiversity," and like many other influential organisms and physical processes before us, we have changed our planet in a geological heartbeat. Geologists now recognize that the planet has entered a new geological era—the Anthropocene—characterized by the human-caused signal of mass extinction to be immortalized in the future fossil record.[2] That is not to say that the extinction of species is abnormal in any way—over 99% of all species that have ever existed on Earth since the dawn of time are now extinct.[3] Put another way, evolution—the process by which species diverge and change into new ones—keeps only slightly ahead of extinction *on average* over geological time.

But averages can be misleading. Extinction tends to come in waves that paleontologists call "mass extinction events." Prior to the Anthropocene, there were five recognized mass extinction events since the great Cambrian explosion of life (about 500 million years ago), of which the Permian extinction (250 million years ago) was the most devastating—around 95% of all species on Earth at that time disappeared.[4] The most infamous mass extinction happened about 65 mil-

*Parts of this chapter were originally published in CJA Bradshaw, "Biowealth: All Creatures Great and Small," in *The Curious Country*, ed. L. Dayton, 30–33 (Canberra, Australia: Office of the Chief Scientist of Australia, ANU E Press, 2013) (with permission to reproduce).

lion years ago at the end of the Cretaceous when a giant asteroid struck Earth and killed off many plants and animals, including most of the dinosaurs (except for what are now birds). The Anthropocene is now characterized by species extinction rates that exceed the background rate—that is, the rate occurring between mass events—by, conservatively, orders of magnitude. Of course, scientists debate the true inflation factor because of the difficulty of observing extinctions in the past and today, and the uncertainty surrounding the true number of species. These issues aside, it is clear enough that we are now losing biodiversity at an alarming rate.[5] For example, a 2014 study by the World Wildlife Fund estimated that Earth has lost around half of its wild animals (total abundance of mammals, birds, reptiles, amphibians, and fish) in the past forty years.[6]

It is easy to be impressed when considering the variety of life on Earth. Conservative estimates place the number of species, not counting multitudes of viruses and bacteria, in different groups living today at more than 4 million protists (eukaryotic, mostly single-celled microorganisms), 75,000–300,000 helminth (worm) parasites, 1.5 million fungi, 320,000 plants, 4–6 million arthropods (insects and the like), 30,000 fishes, 6,500 amphibians, 10,000 reptiles, 10,000 birds, and over 5,000 mammals. However, it might surprise you to learn that biologists described an average of 17,500 new species *every year* over the past decade[7]—we are clearly still a long way from knowing the majority of the species on our own planet. While scientists are nonetheless confident that they have inventoried most of the larger species such as mammals and birds, estimates of the number of smaller, more cryptic species are highly uncertain. In fact, the total number of species on the land and in the oceans worldwide has been estimated to range from only a few million to several hundred million species. Both extremes seem unlikely, with more recent estimates placing the number of non-protists conservatively between the 5 to 9 million mark.[8] In contrast, the collective efforts of all biologists—living or dead—have thus far described only about 1.5 million species in total.[9]

The term *biodiversity* itself is a rather abstract concept. The simplest way of estimating it is just to count the number of species within a given area, but this belies its complexity and the inherent difficulty of defining species in different taxonomic groups and circumstances.[10] Biodiversity includes, among many other things, genetic diversity, ecological function, and the way in which species composition changes over the landscape. Simply adding up the number of species in a given area therefore ignores important things like endemism (species found nowhere else), rarity, genetic variation (how different one population's DNA is from another), evolutionary potential (the ability to evolve in the face of change), and resilience (how easily a species can resist damage in the face of environmental challenges and recover from them). It is little wonder that the average person has difficulty grasping the importance and complexity of biodiversity, especially considering people's increasingly nature-disconnected lifestyles.

Another important aspect of biodiversity is how much of it is disappearing and at what rate. Extinction might appear intuitive to the untrained, because it ultimately involves comparing a time when a species was present to another when it is no longer. Unfortunately, it is not that straightforward. Even the date of the infamous dodo extinction is uncertain, with claims that the species survived another thirty years beyond its last sighting.[11] The problem lies in the fact that as the number of individuals in a population declines, it becomes more difficult to detect remaining individuals, especially in the case of already rare and cryptic species. For example, would anyone even notice if a rare species of underground fungus went extinct? The answer is only if someone had already been documenting its distribution and decline. Expand that to the millions of species on the planet, combined with the uncertainty associated with that number itself, and you begin to understand why our estimates of extinction rates are highly imprecise. However, knowing the exact rates is unimportant. Habitat destruction is the main driver of extinctions—populations and species cannot exist without appropriate habitats. Species are disappearing before our very eyes. Just as we can tell that a beach is eroding rapidly without knowing how many grains of sand are lost per hour, we can determine that biodiversity is declining quickly without such detailed information.

Another concept of species loss that escapes most people's understanding is what is known as "extinction synergies." This is basically the phenomenon that the whole is a lot worse than the sum of its parts. In other words, a species rarely, if ever, goes extinct from a single driver. The question on most curious minds with respect to the reason

Giant redwood trees, *Sequoia sempervirens*, of California's Sierra Nevada mountains. Photo by Corey J. A. Bradshaw.

why a species is no longer with us is typically "What caused it?" However, the synergies concept implies that it is never just one thing—instead, it is an interactive combination of one, two, three, or more punches that knocks a species to the extinction canvas.[12] Consider a lush and intact forest covering a vast area that then succumbs to some human development—let's assume that half of that area is cut down by logging and then turned into cropland. Ecological theory predicts and much empirical evidence demonstrates that because of the positive, nonlinear relationship between area and the number of species it can contain, losing about half of the forest will result in a 20% loss of the species in the original forest (the rate of loss gets worse the greater the proportion of forest that is destroyed). If this were the only form of disturbance—loss of forest area—that the system experienced, it probably would equilibrate at this lower number of species.

The problem is, however, that the loss of area is not the only consideration. Most habitat degradation does not happen in single, contiguous blocks; instead, it tends to happen in a piecemeal fashion that "fragments" the habitat into patches of various size within a "matrix" that is mostly inhospitable to the species living inside the patches. Think of our forest example again—consider small farms interspersed among the greater forest area, small towns, roads, and golf courses. The result is a patchwork of forest fragments within a matrix of land uses in which most forest-dwelling species cannot live. The total area of those patches might still add up to one-half of the original forest, but they no longer make up a single joined area. As such, species with poor dispersal cannot move among the patches, so populations of those species tend to die out in the smallest fragments. This leads to a much greater overall extinction rate throughout the region than would be predicted by our theoretical area-based models.

Other problems make the issue even worse. Roads tend to attract people, and people tend to do things like cut down more trees (for firewood, a place to live), light fires to cook their food, and hunt animals. As the road network pushes farther and farther into the heart of the once-isolated forest, these secondary "drivers" push more and more species over the edge of the extinction precipice.[13] Other subtler changes can put even more pressure on the surviving species. Fragments of forest often experience radically different micro-climatic conditions at their edges. The wind tends to blow faster and dry out these edges, and invading species introduced by humans can penetrate into the fragments by traversing these matrix-forest boundaries. With drying can come a higher risk of burning, leading at the very least to a change in the plant and animal species that can live in this transition zone. These so-called "edge effects" can sometimes penetrate several kilometers or more into the fragment's interior, a major factor in changing the habitat, especially in tropical rain forests. An overall loss of forest can even reduce the amount of rain that falls in a given year[14] and how violently that rain is drained away in floods after it falls.[15]

Comfortably ensconced in our lavish Australian or American homes near to every possible convenience, it is at times difficult to contem-

Threatened in most parts of their range, koala populations are declining due to habitat loss, dog attacks, disease, and inbreeding. This koala decided that C.J.A.B.'s garden trellis was a good survey spot for the day. Photo by Corey J. A. Bradshaw.

plate why biodiversity is at all important to our well-being. It is not an exaggeration to state that our lives depend absolutely on other species. Some relatively simple facts suffice to illustrate this. Consider the very air we breathe—nearly all the oxygen in the atmosphere is produced by plants, and much of that by marine algae—yet we treat our oceans like giant toilets and happily cut down forest blocks worldwide every year that collectively equal the size of Tasmania or West Virginia. Fortunately, our hardworking green friends have already produced so much oxygen that we need not fear suffocation in the near future! On the other hand, the world is now faced with centuries of tumultuous climate disruption from our industrial emissions, yet over 90% of the world's carbon is stored in forests and oceans.[16] In other words, more forests equals less carbon in the atmosphere, and slower, less intense climate change. Much of the food grown to feed the 7 billion-strong human population is pollinated by a wide array of animals, but most of that is done by a single species—the honeybee. Yet populations of many species of bees around the world are crashing because of forest loss, fragmentation, and human overuse of pesticides. Without animal pollination, many crops that make important contributions to the quantity and nutritional quality of the human diet cannot be grown.

It is a hard sell to most people, who can barely appreciate how the supermarket is supplied with foodstuffs from traditional farming, that millions of tiny species such as bacteria, fungi, algae, worms, and insects make our lives possible. Even the air you breathe, the coffee you drink, and the toast you munch exist courtesy of three tiny organisms and the communities of which they are part. *Prochlorococcus* should be a name as well known as "oxygen" because, although a microscopic bacterium, it dominates the ocean plankton that generates over half the oxygen in the air. It is one of Earth's most abundant organisms and plays just as important a part as forests as the "lungs" of the planet. *Trigona* is a small stingless bee that pollinates coffee and about ninety other crops worth billions of dollars worldwide. It is especially important today as it steps up to replace the common honeybee, which, as we indicated, is declining worldwide. The farmer's best friend, *Trichoderma*, is a soil fungus that converts one year's crop stubble into the nutrients required in next year's crop such as the wheat in your toast. Every handful of healthy soil contains billions of valuable soil microbes vital to agricultural production.

Pest control is another crucial but basically unheralded ecosystem service that humanity cannot do without. Pests, including plant-eating insects, devour between a quarter and a half of all the crops produced annually worldwide. In the early part of this century, it was estimated that pesticide production was a $32 billion industry, and over 2.3 billion kilograms (5.1 billion pounds) of pesticides were applied globally to crops each year.[17] Changing public attitudes and problems with the toxicity of pesticides and the ability of pests to become resistant to them are now increasingly promoting a shift to organic farming, in which other techniques for pest control are substituted for chemical pesticides. One of these is insect control by birds, but it is difficult now to predict the magnitude of their potential contribution. For example, it is likely that insectivorous (insect-eating) forest birds visiting crop fields adjacent to tropical forests help suppress agricultural pests; but this question has been poorly studied. More research is needed to determine whether the presence of insectivorous birds broadly indicates healthy agricultural ecosystems. In the long run, their free pest-control services will likely gain prominence as the world's toxification problems become ever more obvious and potentially lead to greater restrictions on the use of pesticides and other poisons that are now being detected in wildlife from pole to pole, the snows of Mount Everest, and, of course, in virtually every human being's blood.

Some studies already reveal that insectivorous birds do play an important role in controlling outbreaks of herbivorous insects in agroforests (stands of trees managed for production of certain crops). Coffee is a commercially important, small understory shrub that traditionally has been grown in shaded plantations—that is, among forest trees. Treeless coffee farms have, however, sprung up across the tropics to produce higher yields faster. Shaded coffee plantations harbor a greater diversity and abundance of birds than do the sunny plantations. It might be expected that many insectivorous birds would control coffee pests effectively in shaded plantations. Indeed, when ecologists simulated a pest outbreak by placing caterpillars in coffee plantations in Chiapas, Mexico, birds quickly came to rescue, consuming the caterpillars. But this prompt pest-predation response occurred only on the farm with high plant diversity and not in the sun-exposed coffee farm. In some other studies, however, the amount of insect consumption by insectivorous birds did not differ between shade and sun coffee plantations. While more research is needed to assess how land-management practices affect the pest-control services provided by birds, it is clear that preserving bird diversity can be essential for maintaining control of insect pests.

The same can be said for the graceful avian insectivores that stay aloft almost all the time. Sadly, many populations of swallows and swifts are now shrinking. Declines of these winged insect vacuum cleaners appear to be widespread, including, for example, a large reduction of whip-poor-wills in Canada—but again the needed studies of the decay of human life-support systems are mostly lacking. Other aerial insectivores such as barn swallows (the most widespread of them) and bank swallows also appear to be in trouble. What seems certain is that such birds mop up populations of mosquitoes, which are not just pests but are important carriers of disease, including some of the worst plagues that afflict humanity—malaria, yellow fever, and dengue. In North America, some of the dangerous diseases they spread are West Nile virus and viral encephalitis. Thus the woes of these once-abundant birds might translate into woes for us. Birds appear to reduce insect abundance most in places where the diversity of bird species is highest.

Bats are intriguing and largely unappreciated creatures. They do not, as folktales claim, make nests in peoples' hair, but they do play essential roles in maintaining human well-being. For instance, huge bat colonies in northern Mexico and the southern United States, comprising up to 20 million individuals, can collectively eat 40 tons of insects every night, many of them insect pests. The potential value of this ecosystem service has been estimated at billions of dollars annually. In a Mexican shaded coffee plantation, researchers have found that bats reduced populations of arthropods (insects, spiders, mites) by more than 80%. Other studies suggest that bats might harvest insects even more efficiently than birds. Consequently, the decline in bat populations worldwide will certainly jeopardize the vital pest-control ecosystem service, not only for coffee growers, but other agriculturalists as well. In the United States, insectivorous bats annually consume some US$750,000 worth of cotton pests in just eight counties of Texas. Across the country, bat-mediated control of insect pests is valued at several billion dollars annually. That free service from nature is now jeopardized by the rapid spread of a new, little-understood, fungal white-nose disease that has killed millions of bats as they hibernated in northeastern North America. It is now moving into the central states, and it is thought that the main vector is an overpopulated mammal species, *Homo sapiens.*

Bats are also important because they pollinate many plants of economic or ecological value and disperse the seeds of many tropical trees. Without bats, many species of plants, directly or indirectly important to humanity, would become extinct for lack of pollination and/or seed dispersal. Unfortunately, many bat species are threatened by climate disruption, impacts from introduced snakes, building of wind turbines, destruction of bat refuges and habitats, or by deliberate overkilling—sometimes using fire to kill all the bats in a cave.

Other agricultural systems benefit from birds and bats as well. Oil palm cultivation has been rapidly expanding, seriously endangering native biodiversity, in the past few decades in the tropics, particularly in Southeast Asia. Palm oil has many uses; in addition to its employment in cooking, the oil is used in soaps, candles, cosmetics, and increasingly as biodiesel. Much of the demand for palm oil is generated in high-income nations like Australia and the United States. Globally, the land under oil palm cultivation has tripled since 1961. More than half of the oil palm expansion in Malaysia and Indonesia has occurred at the expense of primary moist forests, devastating native plants and animals. The conversion of forests to oil palm plantations results in the near disappearance of the majority of forest birds, mammals, butterflies, and other animals. The current rapid spread of oil palm plantations across the tropics therefore spells doom for much of native biodiversity, great increases in emissions of global-warming greenhouse gases, but huge profits for oil palm barons and their paid propagandists. Think of that the next time you bite into a chocolate bar or eat a bag of chips.

It would be appropriate to refer to oil palms as "blood palms" because expansion of their plantations threatens the livelihoods of poor rural people. Oil palm expansion in Indonesia, for example, has resulted in human rights violations through land grabbing and deforestation. Even in government-run programs, local farmers are asked to give up 10 hectares (25 acres) of their land to oil palm companies and in return receive rights to only 2 hectares of the land under oil palm cultivation. Oil palm plantations are also considered a threat to both cultural heritage and human health. The latter is due to the loss of land to cultivate food crops, thus putting local food security at risk. In Southeast Asia in particular, this kind of land grabbing and generation of poor, landless peasants could add to the already burdensome problem of refugee management in neighboring Australia.

Oil palm plantations sometimes suffer from insect pest outbreaks, which can cost oil palm companies a lot of money. Naturally, many of these companies frequently resort to pesticides, despite the likely harm to the environment and to human health. Other approaches such as planting beneficial plants that attract predators and parasitoids of oil palm pests have also been attempted, but with limited success. In Borneo, however, when insectivorous birds were present in oil palm plantations, foliage damage was reduced by about a quarter compared to areas where birds were largely absent. Defoliation lowers fruit yields, so this service by birds enhances the production of the oil. Oil palm companies should take steps to attract insectivorous birds and the pest-control service they supply. But the oil palm planters have no such farsighted ideas. The greed of oil palm promoters is difficult to overestimate, and they run a worldwide campaign of disinformation.[18]

In addition to pest control, other comparable examples of ecosystem services abound. These include climate control, providing fresh water, crop pollination, supplying food, pharmaceuticals, air purification, waste disposal, flood control, and many others. Even the much-maligned shark is an essential ecosystem engineer. Wherever shark populations are abundant and diverse, reefs are healthier, fish populations are larger, and even the water is clearer.[19] This is because large sharks impose a top-down pressure on smaller predators, thus limiting the latter's intake of other fish species. Removing the biggest predators means that smaller predators increase, which then quickly eat out other species that keep things like algae in check. The overall effect is a biologically poorer system now more prone to further degradation. We should not be too afraid of sharks either—even in Australia the probability of being attacked by a shark is several hundred times lower than just drowning during a quick swim or surf.[20]

If one considers the totality of all these different interactions, dependencies, and functions—the subject of the scientific discipline we call "ecology"—the logical conclusion is that, in relation to human well-being, all of biodiversity can be considered under the umbrella of biowealth. This concept encapsulates the two most important elements of biodiversity from a human perspective: (*i*) that *diversity* is an essential requirement for life; without all, or at least most, of these species, we inevitably lose important functions like plant pollination, water purification, and carbon sequestration; and (*ii*) this diversity provides humanity—largely free of charge—the elements essential for our own survival. For instance, if there were no natural pest controls, humanity would be out of the business of high-yield agriculture, and many billions would starve. Without biodiversity, we are poor; with it, we are

A mischievous sea otter in Prince William Sound, Alaska. Photo by Corey J. A. Bradshaw.

biowealthy. Biodiversity, of course, feeds the soul as well as the body. For at least hundreds of millions of people, the beauty, behavior, and intricacy of other organisms are a source of joy, inspiration, fascination, and relaxation. Those people relate to our only-known companions in this vast universe.

With this in mind, it is clear that we should factor at least some elements of biowealth into our economic plans. The problem is that their economic value is difficult to calculate. How do you value atmospheric oxygen? We cannot trade it on the stock market, although you can sell it in bottled form. Perhaps the insects that pollinate crops and the fungi that fertilize farm soils could be assessed as some fraction of the billions of dollars generated by global agriculture. Similar credit could be assigned to the bugs, birds, bats, and other friends who control the pests that would otherwise devour large portions of the crops we grow. The same situation applies to the food chains that generate our fisheries and the soil communities upon which forestry is based. Further, many minuscule organisms help regulate global carbon and provide the inspiration or blueprints for an astonishing range of industrial products, from cold-wash enzymes to supersensitive fire alarms.

But these organisms are generally unknown to the industries that

One of Australia's better-known and more dangerous species, the estuarine crocodile *Crocodylus porosus*. Photo by Iain Field.

depend on them—or their investors for that matter. Entire sectors of the economy apparently rely on components (populations and species) about which almost nobody in the business has a clue. Instead, most people subconsciously hope they are out there and functioning. What we do know is that these millions of species remain a scientific and policy Cinderella. This is because conservationists—whether government, nongovernment, or, regrettably, many scientists—are almost completely preoccupied by popular charismatic species that make up at most a mere 5% of species on Earth. In this context, biowealth is not solely the concern of conservationists; it is in fact at the core of the economy and affects every one of us every day.

So consider the crocodiles, the sharks, the snakes, the small and the squirmy, the smelly, the slimy and scaly; consider the fanged and the hairy, the ugly and the cute alike. The more we degrade this mind-blowing diversity of evolved life and all its interactions on our only home, the more we expose ourselves to the ravages of a universe that is inherently hostile to life. Let's embrace, protect, and cherish our biowealth, at least so that our children can live happy and prosperous lives. We could even incorporate the concept better into our daily lives by regularly reporting the state of our nation's biowealth alongside economic, sport, and stock market indices. Only then will society be cognizant of, and perhaps stimulated to improve, the state of our one-and-only life-support system.

5 Liquidated Assets

Living wild species are like a library of books still unread. Our heedless destruction of them is akin to burning the library without ever having read its books.

CONGRESSMAN JOHN D. DINGELL,
in *Balancing on the Brink of Extinction* (1991)[1]

Up until now we have focused on the history of change in two of Britain's better-known ex-colonies. Damage has certainly been done, but how much? It might surprise many Americans, and probably even more Australians, just how ecologically compromised their respective countries are. Neither do most people likely appreciate the effect of such damage on their own livelihoods and long-term prospects. We still tend to think of our countries as "new," only recently colonized by Europeans and, therefore, relatively "wild." Even the term "wilderness" was coined in the United States—a seemingly limitless landscape filled with European notions of dangerous creatures and harsh climates that were generally incompatible with the Old World ideal of tranquil countrysides devoid of nature's imminent danger. While true that the Old World is decidedly tamer than her faraway offspring, her wayward children have been particularly naughty. The "wilderness" has more or less disappeared.

As we described previously, Australia was far from a pristine place when the first Europeans settled its east coast in the late eighteenth century—the Aborigines had done a marvelous job of transforming the forests and savannas by the clever use of fire and spear. Although renowned globally for its "Red Centre"—a hot and unforgiving land of dust, saltbush, poisonous snakes, and funky lizards—Australia still had about 30% of its land area covered by forests[2] when Europeans arrived. It comes as no surprise that most of these forests fringe the wetter, coastal regions of the continent, and that these areas also contained some of Australia's most fertile soils. As such, the agriculturally driven Europeans first cleared these wetter and most fertile areas in the country as the sheep and wheat industries expanded in a wave across the southeast, eventually reaching the remote western fringes of the continent in southwestern Western Australia. By the 1980s, approximately 38% of Australia's forests had been severely modified by clearing, and today Australia has one of the world's lowest total areas of remaining closed forests[3] (4% of its land area). Australia's native forests now cover 147 million hectares (363 million acres), or only 19% of its total land area.[4] Today approximately 15% of the continent has been severely modified by intensive land use (crops, cities), with less-modified agricultural areas dominated by cattle grazing zones covering around 43% of the country, and "improved" pastures covering around 10%.

Even many of Australia's remaining forests are severely degraded; about 50% of them have been completely cleared or greatly modified at one time or another, with over 80% of eucalypt forests in particular having been altered in some way by humans. Even for those eucalypt forests now under some type of protection, over 50% of those have been logged at some point in the past 200 years. Much of the remaining forest cover is severely fragmented (converted to small forest patches), especially in southeastern Australia, with roads, urban development, agriculture, and (alien) pine plantations isolating existing fragments to the point that much of their biowealth is severely compromised.[5]

One of the saddest periods of Australia's love affair with vegetation clearance for agriculture is the case of Western Australia's "wheatbelt." Separated from the rest of Australia by harsh deserts and arid lands, the biodiversity of Western Australia has evolved over 4,000 endemic (found nowhere else) plant and 100 vertebrate species.[6] So unique are its species that the southwest of the state is now one of Australia's only two* "Biodiversity Hotspots,"[7] a dubious distinction because it signifies an area of the planet with an extraordinary number of species found nowhere else, but that has simultaneously expe-

*Australia's other Biodiversity Hotspot, "Forests of East Australia," was named the thirty-fifth International Biodiversity Hotspot in 2011.

rienced an "exceptional" loss of habitat in recent years. Exceptional it has been—well over 90% of this Hotspot's 310,000-square-kilometer (120,000-square-mile) area of primary vegetation has been cleared, mainly for the expansion of wheat cultivation. While not unlike other regions of Australia with respect to the underlying causes, this loss is all the more tragic because it happened mainly in the mid-twentieth century—54% of all land developed for agriculture was cleared from 1945 to 1982, and with clearing of between 30,000 and 60,000 hectares (74,000 to 148,000 acres) per year continuing well into the late 1980s.

But this was not your typical removal of vegetation by private landholders keen to make a dollar—it was a multi-decade policy of the Western Australian government to promote the active clearing of the land, much as the New South Welsh embraced deforestation in the nineteenth century. "Clear it or lose it" was the motto of the day, and clear with fervor they did. A common image of these times was the enormous, house-size "hiball" pulled along by 200 meters (656 feet) of chain attached to bulldozers—never a more effective bush razor was invented. The immense push by the government to advance the country's agricultural capacity following the Second World War, a technological leap in postwar machinery, and the proud psyche of the Australian farming tradition created a lust for forest clearing that has been unrivaled any time before or since in the country's history.

But there were severe consequences for that lust (there usually are), and not just in the cumulative devastation of the region's unique flora and fauna. The rapid and expansive clearing brutally upset the delicate water balance in the soils of that ancient landscape. Much of inland Australia has been inundated by seawater at various times over the last several hundred million years, a product of plate tectonics and large shifts in global temperatures leading to impressive sea-level changes that periodically flooded this relatively low-altitude landscape. Additionally, oceanic winds carrying sea salt have transported huge quantities of salt to the land, and the erosion of salt-bearing parent rocks over millions of years all mean that Australian soils have an unusually high salt content. When the land is disturbed and irrigated over large areas, salt migrates to the surface, effectively poisoning the soil for plant growth. Widespread throughout Australia now, this salinity problem affects well over 3 million hectares (7 million acres) of arable land in the south, with the majority of that area in Western Aus-

The massive "hiball" dragged between bulldozers to level vegetation. Photographer: Frank Hurley. Soldier settlement, Rocky Gully, Western Australia (between 1910 and 1962). Permission to reproduce courtesy of the National Library of Australia.

tralia and New South Wales. The economic implications are measured in the billions of dollars of lost agricultural opportunity, with the prospect of much more area of saline deserts forming in Australia's arable lands over the coming decades. Oops.

If deforestation, extinctions, and salt poisoning are not enough to depress you, then consider the additional fact that Australian deforestation has actually changed the climate—for the worse. Clearing large swathes of vegetation really has two main effects on climate. First, the mere act of cutting down trees releases huge amounts of carbon dioxide and other greenhouse gases into the atmosphere, which in turn heat up the globe. In 1980 deforestation in Australia contributed about 28 million tonnes of carbon that year and represented about 22% of the country's total annual greenhouse gas emissions by 1995. Loss of forest also means a loss of carbon uptake potential, for as forests grow, they extract large amounts of carbon dioxide from the atmosphere with which they create complex sugars during photosynthesis. It has been estimated that if the eucalypt forests of Queensland to Tasmania

Dryland salinity causes trees to die and induces serious erosion (Charters Towers, Queensland, 1998). Photo by Willem van Aken; reproduced with permission from CSIRO.

were still intact, they would sequester about 33 billion tonnes (36 billion tons) of carbon, much of it in the soil, each year. Since over half of these are now gone, Australia has lost a great way to reduce its overall greenhouse gas emissions (something that should not be underestimated, for Australia has one of the highest per capita national greenhouse gas emissions rates in the world). Second, removing vegetation changes the local climate by disrupting how plants respire and cycle water, how soils retain moisture, and how the sun's energy is dissipated when it hits the surface. There is ample evidence now, especially for Western Australia, that the extent of mass historical deforestation has increased local temperatures and reduced rainfall enough to cause long-term droughts that further reduce the region's agricultural capacity.[8] Double oops.

But Australia did not learn its lesson quickly, and it clearly has a long way to go yet on the path to environmental enlightenment. Queensland, Australia's largest state (nearly 1.9 million square kilometers, or 734,000 square miles)—and originally home to the continent's largest extent of tropical, subtropical, and warm temperate forests—was quick to take over as the country's worst deforester. In many ways,

Queensland's vegetation-clearance patterns are remarkable in that most deforestation occurred in the last fifty years, mainly due to the expansion of the cattle industry. Under the ultra-conservative and corrupt state Premier Sir Johannes Bjelke-Petersen, who ruled like an iron-fisted dictator from 1968–87, rapid and poorly planned developments were the hallmarks of his environmental tyranny. "Sir Joh" had a largely agricultural background and specialized in land clearing (he was even possibly credited with having developed the "hiball" clearing system). As Premier, though, he was far more influential on other aspects of environmental degradation—he was instrumental in the development of the Gold Coast, which has become an eyesore of high-rise buildings perched precariously just above the beaches' high-tide mark, gaudy casinos, and tacky amusement parks along some of the country's once most beautiful coastlines, which were greatly admired by P.R.E. when he flew a light aircraft through the Hinchinbrook Passage in 1965. As a result of this Disneyfication, perhaps Australia should consider changing the state's name from "Queensland" to "Queensworld." Sir Joh also actively promoted the construction of several dams, including the Hinze, Wivenhoe, and Burdekin; coal-fired power plants; and many of the country's largest freeways. His government was also impervious to calls to limit clearing and, in most cases, actively promoted the reckless agricultural "development" of the state. In fact, it was not until the mid-2000s and nearly twenty years after his departure from politics, that any form of restriction on forest clearing was imposed in Queensland. From the period of 1981–2000, Queensland was dubbed a global "deforestation hotspot."[9]

Although the deforestation battle in Australia seemed mostly over by 2010, the war to save what remains of its biodiversity is only beginning. After squandering what little forest cover it had, Australia is struggling (and largely failing) to limit the damage to its biodiversity. It bears reminding that most deforestation has been done in the name of agriculture and, therefore, in the most productive areas of the country. Generally speaking, more "productive" regions (i.e., good soils, sufficient rainfall, relatively benign climates) have more species of plants and animals, so Australian plants and wildlife have already lost the best real estate to farming. There is also the issue of extinction "lags," which basically means that it takes sometimes decades, or even centuries, for some of these "walking dead" species to disappear after

Despite most deforestation occurring previously, clearing bush is still ongoing in many parts of Australia. This photo is of recently cleared bush in northern Queensland in 2009. Photo by Corey J. A. Bradshaw.

their habitats have experienced loss or degradation. In other words, we have yet to experience the full scale of extinctions and biodiversity decline resulting from deforestation that occurred decades ago.

Curtailed vegetation clearing belies a more insidious threat to forest species—the "health" of the remaining forest landscape. Unfortunately, Australia's remaining forests are highly fragmented and highly "disturbed" by encroaching weedy species or partial logging, or otherwise ecologically "compromised" by, for example, being overrun with feral animals. Even twenty years ago, more than 80% of Australia's iconic eucalypt forests had been modified in some way by humans. This extends to protected areas—more than 50% of forests under any form of legislated protection, such as reserves and national parks, have been logged at some point in the past 200 years.

So why is fragmentation and degradation so bad, even if from an aerial view forests seem to be intact and expansive? One of the best understood phenomena in ecology is that as a "patch" of habitat—be that a forest, a coral reef, or a desert oasis—gets smaller, the fewer species (and individuals making up each species) it can hold. So when we reduce the size of a forest or cut it up into smaller fragments, each one eventually has fewer species than the uncorrupted total would have had. Isolation is another nefarious component of this evil equation—the farther away a patch is from others, the fewer species it can hold. To pilfer from an old saying, this is because no habitat fragment is an island, in that small populations cannot usually sustain themselves solely by their own reproduction. Most isolated populations need to be "rescued" periodically by immigrants from other populations. The farther away you are, the fewer immigrants you get, and the higher the chance your population will go extinct—this has even happened to humans: for example, Kangaroo Island off the coast of South Australia was once occupied by Aborigines, but they were no longer present when the Europeans arrived.

So most of Australia's remaining forests are dying a death by a thousand cuts. For example, in the Lockyer Valley catchment of Queensland, the average size of forest fragments decreased by 54% from 1973 to 1997, whereas in the Kellerberrin region of Western Australia's wheatbelt, 93% of its native vegetation has been removed since 1940, resulting now in most of its habitat fragments being less than 20 hectares (50 acres) in size. The Mount Lofty Ranges of South Australia have one of the saddest histories of deforestation and degradation in the country, and there is little sign that things are improving.[10] Today only 10% of its original vegetation remains, and most of it is highly fragmented. By the 1990s, the mean size of its 4,000 remaining native habitat patches was only 13 hectares (32 acres). For its native bird species, this is about a hundred times smaller than it should be to prevent further extinctions. Indeed, the region has lost at least 132 species of native animals and plants, and has been invaded by over 648 alien species (mostly weeds). Some scientists have even suggested that this region of southern Australia should be treated as the country's ecosystem "canary"—when that little bird stops singing, it is probably game over for the bulk of Australia's temperate forest biodiversity.

Australia is also overrun with feral animals. Previously, we only mentioned the herbivores—while their damage is extensive, the worst invasive menaces are decidedly smaller and sharper—foxes and cats. For North Americans, the concept of the "feral" cat might seem anathema. How could the family pet ("moggy" in Australian parlance)—that cute and cuddly ball of fluff that sleeps all day—be a menace to biodiversity? Any cat owner will probably admit, though, that Fluffy is an amazing potential killer when given the chance. How many mangled and partially eaten bird carcasses have been deposited on cat-owner doormats around the world? According to US Humane Society statistics, there are over 86 million domestic cats in the United States, with about one-third of all American households owning at least one cat. There are probably another 30 million feral cats roaming the country. Together, this means a massive toll on biodiversity, for cats dine regularly on native birds, other small mammals, reptiles, and amphibians. Considering the popularity of bird-watching, bird hunting, and the pet trade in the United States, this equates to a loss of at least US$17 billion per year *just from cat predation on birds.*[11] Not surprisingly, both Australia and the United States are following in the tradition of their mother country—domestic cats are major predators in England too.

Australians also love their house cats (there are about 2.5 million in Australia), but the problem is especially serious there because Australian ecosystems never evolved with feline predators. In North America there are lynx, bobcats, mountain lions, some ocelots, and occasionally jaguars. There are also many other small to midsize efficient mammalian predators like weasels, mink, marten, wolverines, coyotes, and wolves. Australia never had predators quite like this menagerie of killers, so when domestic cats started becoming feral, the impact on Australian animals was catastrophic. Today there are probably well over 12 million feral cats across Australia. Dr. Tony Peacock, former head of the Invasive Animals Cooperative Research Centre in Canberra, once used a gut-wrenching analogy to explain their impact on Australian native animals.

> If you take a line of livestock-transport trucks placed bumper-to-bumper from Sydney to Grafton [over 800 kilometers, or 500 miles], and fill them with native animals, such as lizards, penguins, possums and the like—this is how much is eaten *each year* by feral cats in Australia.

This sobering analogy explains why, at least in part, Australia leads the world in modern mammal extinctions. The other big "part" is loss of habitat, of course. Australia has permanently lost to the vaults of geological time twenty-three species of mammal,[12] such as various rat kangaroos, wallabies, bandicoots, potoroos, bettongs, bilbies, and hopping mice. Another hundred species are highly threatened with extinction for largely the same reason.

One area that has been the subject to much hand-wringing in this regard is Australia's largest national park—Kakadu National Park—encompassing nearly 20,000 square kilometers (7,700 square miles). There over the last twenty years, almost every remaining native small mammal population is in decline—at times leaving only 5% of the original population—and even to the point of local extinctions in over 50% of all monitored sites. Of course, cats play a large role in these declines,[13] but they are not the sole cause. Inappropriate fire management contributes—northern Australians love to burn, with the excuse that the bush will burn anyway, only hotter and faster if people do not prescribe-burn. The problem is, however, that we can burn much larger areas more frequently than we could have previously given technological breakthroughs like helicopters and drip-torches. There is also ineffective overall intervention to control invasive species and weeds that can make fires worse. Remember—this is in a *national park*, which is supposed to buffer the species within from exactly these kinds of pressures. Worldwide, protected areas are failing for exactly the same reasons[14]—not enough attention or money applied to managing the problems, and external forces such as growth of surrounding human populations, deforestation, hunting, and runaway fires penetrating reserves from the outside. In 2009 a visiting South African ecologist (Professor William Bond) commented* on the extinction crisis unfolding in northern Australia:

> When you arrive in Australia, you are bombarded with slogans of sporting victories, great food, and fantastic wine, but no one tells you of the biodiversity tragedy that has, and is continuing, to happen in Australia. In Africa, we have managed to convince people to conserve elephants that destroy their crops and kill their families—why can't Australians realize they are destroying their very heritage? What are you doing to stop the carnage?

Part of the problem is that in reality, few of Australia's 500 or so "national" parks are national at all. Covering 28 million hectares (108,000

*Paraphrased.

square miles) of land, or about 3.6% of Australia's land surface, you could be forgiven for thinking the country was in good shape in terms of protection. Of those 500 "national" parks, only 6 are managed by the Commonwealth (federal) government. For marine parks, it is a little more: 61 of the more than 130 are managed primarily by the Commonwealth. This means that most of Australia's important biowealth refuges are managed exclusively by state and territory governments.

In a world of perfect governance, this would not matter. But Australia is experiencing the rapid "relaxation" by many state governments of laws designed to protect its "national" and marine parks. One bad decision resulting in a major ecosystem disturbance in a national park can take decades, if not hundreds of years, to heal. Ecosystems are built of complex interactions of millions of species that take a long time to evolve—they cannot be easily repaired once the damage is done. Within weeks of the newest, conservative federal government taking power in 2013, the states of Queensland, New South Wales, and Victoria rolled back over a century of park protections. What was surprising was that many of Australia's conservation gains in the last few decades (e.g., the Natural Heritage Trust, the National Reserve System, the Environment Protection and Biodiversity Conservation Act, and a national marine reserve network) originated from that very same right-wing party years before.

With six quasi-autonomous states and several territories in Australia, the probability of implementing bad biodiversity policies is considerably higher than if a single central government agency was responsible for their protection. Should the federal government extend its power to veto potentially high-impact activities like logging, grazing, and mining proposed in national parks by the states? As ecologists, we could not agree more with the intent of such a proposal, although the power to veto high-impact activities would not be enough to stem the tide of damage. Vetoes could be legally challenged by the states, and co-operation would depend on which parties were in control at state and federal levels at any one time. Veto power alone could make for expensive and drawn-out court battles. As far as legalization goes, what Australia needs is more concrete federal legislation restricting specific activities like fishing, logging, and grazing.

It is no guarantee, of course, that the Commonwealth would be any better at managing national parks than the states. Kakadu National

Park—Australia's largest and possibly most important national park, and managed by the Commonwealth—is a global conservation embarrassment. Undermined by internal management bickering and insufficient funding, its unique biodiversity has been declining sharply for decades. The best thing the federal government could do would be to allocate a little more money to national parks. More evidence is coming to light that for national parks to maintain healthy ecosystems, they require substantial effort on the part of their managers and adequate funding.[15]

Adequate funding would not drain the national budget dry. A study in 2012 estimated that it would cost up to US$4.76 billion annually to protect the entire planet's threatened species from extinction,[16] or up to US$76 billion annually to protect and maintain protected areas like national parks. To put that into perspective, the cost amounts to less than 20% of the global annual budget spent on soft drinks each year. If we therefore allocated only a little more to finance the management of Australia's national reserve system and a little more for local communities to get involved, we could fix many of the conservation problems its national parks face today. Fortunately, a few independent organizations are taking up some of the slack from Australia's failing national parks system. For example, the Australian Wildlife Conservancy is a ten-year-old institution with two simple goals: (1) establishing as many sanctuaries as possible by acquiring land through partnerships with landholders and (2) managing the threats in those sanctuaries, such as feral predator control. By securing land, applying science, and implementing management, they are succeeding at protecting representative areas of Australian nature and safeguarding many of the endemic species that will otherwise will be lost if they were to rely solely on government-run protected areas.

On the subject of funding, Australia has invested billions of dollars in feral animal and weed control over the past several decades. However, the nation is no closer to solving these problems than it was fifty years ago. One reason is that not a lot of science is used in the process, and crazy schemes supported by influential lobby groups end up winning the day. A case in point is the Australian approach to "managing" the dingo.

The dingo has long evoked fear and loathing in the hearts of Australians. Ever since we learned that it was introduced around 4,000 years ago by Southeast Asian visitors to Australia's northern shores, many Australians have developed the irrational opinion that this sheep-killing, baby-stealing, thylacine- and Tasmanian devil-displacing feral from Asia is a menace that should be eradicated at all costs. But when one looks at the evidence behind these xenophobic attitudes, the image becomes much less clear. Despite some high-profile incidences of attacks on humans,[17] dingoes are perhaps one of the least-dangerous species to people in Australia. The entirely coincidental disappearance of thylacines (Tasmanian tigers) and devils* from mainland Australia when the dingo appeared also ignores that the climate was changing and Aboriginal populations began booming at the same time.[18] In fact, recent computer simulations show much more support for a human- rather than dingo-caused extinction of those two native marsupials.[19]

The biggest problem, though, is the pastoralism industry, and mainly those who insist that the arid rangelands of outback Australia are suitable for sheep farming. To most pastoralists, the dingo is the manifestation of pure evil—a bloodthirsty sheep killer that must be eradicated. So in response to the wolf cries of sheep farmers from Queensland to Western Australia, Australia built the longest fence in the world. Over 5,500 kilometers long (3,420 miles—roughly the distance between Anchorage, Alaska, and New Orleans, Louisiana), it follows a circuitous route from north of Brisbane in the east to near Yalata, South Australia, in the west. Possibly the world's longest human-built structure,† the dingo fence is a monument to predator xenophobia. Its role is controversial, because while it certainly has prevented a large influx of dingoes into southern and eastern Australia, it has also seen, in a situation of lower predation pressure, a proliferation of competing native (kangaroos) and non-native (rabbits) herbivores where dingoes are absent or in low abundance.

While the roughly $10 million it costs each year to maintain the fence is lower than the cited $48 million per year that pastoralists claim to lose to "wild dogs," these costs do not include the labor-intensive and expensive additional poisoning of dingoes that accompanies the fencing. But poisoning is not the answer either. In addition to killing non-

*Tasmanian devils—although now restricted to Tasmania (hence the epithet), they once roamed the entirety of Australia.

†Mostly completed by 1896.

A monument to predator xenophobia: a section of the dingo fence near Roxby Downs, South Australia. Photo by Corey J. A. Bradshaw

target native species, baiting dingoes might in fact result in increased dingo densities due to social breakdown of the pack that creates smaller territories,[20] resulting in increasing attacks on stock, not to mention a higher likelihood of hybridization with feral dogs. Baiting also leads to a higher proportion of juvenile dingoes. These less-efficient predators tend to target cattle calves more than do adult dingoes.

Neither do the "costs" include the unquantifiable negative impacts on Australia's biodiversity. Australia spends millions per year on native species recovery, and how many more are lost from depleted biowealth as a result of their population declines? The blockage of animal migration routes and the death toll imposed by the fence for species like emus and kangaroos are horrific, not to mention the cruelty of entangled animals languishing for days in the outback sun. There is also

Emu migration blocked by a "barrier" fence. When gathering at the fence, these endemic flightless birds emus have been shot, poisoned, or left to starve in the tens of thousands. Photo credit: Western Australia Department of Agriculture and Food, reprinted with permission.

the issue of the fence's effectiveness—today dingoes are penetrating farther and farther south due to camel damage to the fence, and other weaker areas where dingoes can penetrate.

These conservation and welfare issues aside, it turns out that the dingo is instead a sorely underutilized weapon in Australia's arsenal to control feral animals.[21] Wherever scientists have compared areas with and without dingoes, the message is clear—when dingoes are abundant, foxes and cats are not, and native marsupials proliferate.[22] It is called the "mesopredator" effect because the higher-order dingo suppresses, via direct attack and competition, the smaller "mesopredators" (cats and foxes). It highlights the important role of predators in maintaining healthy ecosystems.

There are other advantages to having dingoes around that might not seem obvious. Dingoes reduce herbivore densities, and this can weaken the effects of climate change–induced drought by increasing available plant cover.[23] Dingoes can therefore benefit pastoralists by providing more vegetation to produce stronger, healthier cattle that can resist attack (indeed, dingoes prefer more passive prey such as

A dingo in the Australian outback. Photo credit: Angus McNab.

kangaroos).[24] By eating more kangaroos, dingo predation can actually offset the costs of any livestock losses by freeing up the available grass for cattle, thus making farmers more money.[25] Pest management in Australia—and in much of the United States, for that matter—lacks any sort of integrated approach based on ecological principles. Land managers remove foxes with poison baits and shooting, but then cats increase; cats are removed, but then rabbits increase. Once dingoes are excluded or poisoned, there are more feral herbivore competition problems. This inefficient hopping from one single-species crisis to the next is a waste of money and time. It lacks a long-term vision.

The situation in the United States is similar, where once there were some 5 billion prairie dogs (several kinds of ground squirrel). Ranchers did not like them, imagining that cattle frequently broke their legs in prairie dog holes. So control programs were initiated that annihilated most of their populations and had negative effects on plants and animals that took advantage of the unique habitats they created. But studies of prairie dogs begun in the early 1990s demonstrated that they are instead essential to maintaining their grassland ecosystem's productivity. The extermination of prairie dogs led to an invasion of scrubland vegetation and desertification, largely destroying the land's value for grazing. Control programs were also insanely initiated for coyotes, a major predator of the prairie dogs. Those programs mostly have had no discernible effect, except to illustrate again the foolishness of single-species approaches to real (or imagined) ecosystem problems.

Australia's most common "feral" animals are, however, those most cherished by society—cattle and sheep. Like much of the United States, Australia fancies itself a livestock nation, even though most of its land surface is entirely unsuitable for livestock production. This ecological reality has not stopped Australians from investing billions in the industry, to the point that there are now approximately as many cattle as people (23 million), and about three times as many sheep (68 million). As a result, about three-quarters of Australia is now covered in grazing rangelands.[26] While there are still more cattle in the United States (about 100 million), the ratio of cattle to people there is lower at 1:3. In addition to the massive ecological damage this amount of grazing pressure has produced, Australia's livestock is so abundant that the radiative forcing of the methane emissions (burps and farts) from all those cattle and sheep actually exceeds that from its coal-fired power plants[27] and completely dwarfs the emissions emanating from its other feral herbivores.[28]

Livestock pastoralism in Australia has been implicated in the extinction of at least twenty mammal species and threatens around 25% of the plant species listed as endangered in Australia.[29] It is also becoming more difficult to raise water-thirsty livestock as Australia's rainfall dwindles with climate disruption. But what if Australians switched their protein dependency from feral livestock to native kangaroos? The kangaroo's gut fermentation system produces vastly lower amounts of methane (one of the nasty greenhouse gases responsible for heating up our planet) than livestock rumination, so raising more kangaroos instead of livestock makes good environmental and economic sense under existing and proposed carbon-trading markets.[30] If more Australians replaced at least one meal per week with kangaroo meat, Australia's grazing rangelands would be much better off in the long run. The problem is that most Australians do not eat kangaroo because it is associated with the diet of poor immigrants and ex-convicts from early in the nation's history. Others shun the meat for welfare reasons, despite the clear ethical hypocrisy of choosing to eat beef, lamb, or chicken instead. Despite an expansive and thriving commercial kangaroo-harvesting industry around the nation, to this day most of the meat goes to making pet food—even those supermarkets or butchers providing kangaroo meat for human consumption often display it next to the pet meat section! This is a particular shame, because kangaroo meat is tasty, generally contains less fat than beef or lamb, and any concomitant reduction in cattle or sheep would have enormous benefits for Australia's fragile environment.

Australia's environmental problems do not end with feral pests and livestock. A recent assessment[31] of the state of the Australian environment released some worrying statistics. Between 2000 and 2008, the number of terrestrial bird and mammal species considered to be Extinct, Endangered, or Vulnerable (as assessed by the International Union for the Conservation of Nature Red List of Threatened Species[32]) increased by 14% from 154 to 175 (of which 69 were birds and 106 were mammals). At the end of 2008, just 46% of these were Vulnerable, 37% were more seriously threatened, and the rest of them (18%) were presumed extinct. Between 2000 and 2008, there were increases in the number of both Endangered and Vulnerable species. The rise in species assessed as Endangered was higher (an increase of 33%) than those assessed as Vulnerable (an increase of 7%). This should not come as too much of a surprise given Australia's world-record extinction rate for mammals, but it is telling that the trend is extending to other groups such as birds. Even introduced plants have a new record in Australia—for this first time it was estimated in 2012 that there are more introduced plants than native plants. Many of these are weedy pests that cost the Australian economy some $3.5 billion each year. The trend does not stop there—it extends to the oceans. For fish populations taken in fisheries managed by the Australian government, 16 of the 96 principal species assessed (17%) were deemed "overfished" and/or subject to overfishing.

The signs are no better for Australia's massive marine biowealth. Living on a rather large island with a desert center, Australians are intimately tied to the coast, and Australia is among the most coastline-rich countries in the world. Like all countries with coastlines, it possesses an exclusive economic zone (mainly the area within 200 nautical miles of its coast), within which it has sovereign maritime rights, including fishing. Among one of the largest exclusive economic zones in the world (10 million square kilometers,* or 3.9 million square miles), it encloses the world's largest coral reef system (the Great Barrier Reef) and one of the world's longest fringing coral reefs (Ningaloo). But because of Australia's low population density, remoteness (especially its northern coastline), and harsh monsoon climate, its northern seas are difficult to patrol. As a result, the seas to the north of Australia—including the Java, Timor, and Arafura Seas—are today under ruthless attack of illegal fishing, mainly from Indonesia but often supported by Chinese ventures. The resultant "protein-mining" wave that grew out of the South China Sea in the 1970s has been slowly moving southward and now encroaches on the entirety of the northern Australian exclusive economic zone. The near 90% reduction in shark and ray biomass in the Java Sea[33] since the 1970s is now beginning to herald similar potential fishing crashes in Australia.[34] With inadequate resources to enforce its rights, and a relentless supply of poor Indonesian and other Southeast Asian fishers willing to risk life and freedom to provide for their families, the prospect of catastrophic fish declines is almost certain. Even the relatively well-protected icon of marine Australia, the Great Barrier Reef, is under increasing threat and declining rapidly, mainly from terrestrial runoff and coastal development. And Australia's current anti-environment government has been pressing to permit the dumping of some 3 million tonnes of marine dredge from a coal port facility development on the Barrier Reef. The United Nations Educational, Scientific and Cultural Organization (UNESCO) has even threatened to lift its World Heritage status if things do not improve soon.[35]

The current state of the environment in the United States is a little trickier to describe succinctly, mainly because it is so complex. Its much longer history since initial European colonization compared to Australia and its highly piecemeal settlement pattern mean that it is difficult to generalize. Suffice it to say that most of the "big" land changes happened a lot longer ago than in Australia. Following European settlement, heavy deforestation in the eastern United States started in earnest during the 1600s, and during the eighteenth and nineteenth cen-

*Which is approximately the same size as the US Exclusive Economic Zone, mainly due to the large offshore archipelagos (e.g., Hawai'i) it claims.

turies, regions near the Atlantic coast were almost completely cleared for agriculture, with only small "woodlots" on farms persisting.[36] In Massachusetts alone, nearly 80% of the state's forests was cleared in the single period between 1800 and 1860.[37] As the human-settlement wave progressed to the fertile Ohio Valley, much of that area was deforested by the early nineteenth century. But the history of deforestation in the eastern United States does not follow a pattern of permanent devastation—as the wave of accelerating deforestation passed from the Atlantic coast to the edge of the prairie, forests began to regenerate as people abandoned the easternmost farms. Indeed, by the time that over half of the native forests had been removed from the Ohio Valley, forests in New England had already started a healthy recovery.[38] By the late nineteenth century when little of the original hardwood forests of Minnesota and Wisconsin remained, Ohio forests had started to regenerate. The South also lost most of its forests during that time. The late nineteenth century was therefore the time when the coverage of the eastern forests was at its lowest—roughly half of what it had been at European settlement in 1620.[39]

When considering the total amount of forests and woodlands removed (ignoring regrowth for the moment), deforestation amounts to some 1.4 million square kilometers (540,000 square miles) since 1850 in all of North America (including Canada and Mexico). Even though America's forests have not really changed much in total area coverage since the 1950s,[40] the species composition within them has changed substantially. For example, in Massachusetts, there was a large decline in beech, hemlock, and chestnut trees, and an increase in maple, birch, and cherry following clearing and agricultural abandonment.[41] The age of trees in US forests has also declined on average—there are few old trees left either in the eastern and southern forests, or even in the high-rotation managed (logged) forests of the Pacific Northwest. That has had a massive effect on, among other things, birds such as the ivory-billed woodpecker (now almost certainly extinct) and red-cockaded woodpecker (endangered) that require old-growth forest habitats.

Another legacy of abandoned agriculture is that the regrowth rarely looks like the original forest. For example, the herbaceous understory is typically restricted in the number of species in old agricultural plots compared to virgin forest—and it can take centuries for this diversity to bounce back, sometimes only with a helping hand from human restoration efforts.[42] Even the carbon content in the aboveground plant biomass and soils can take centuries to recover.[43]

Unlike Australia, where urban centers are relatively few, urbanization in the United States is also changing the pattern of forest cover. Cities, towns, and villages cover some 3.5% of the lower forty-eight states, having doubled in area over the past twenty-five years. Metropolitan areas (urban counties) have tripled in extent since 1950, and now built-up areas cover roughly a whopping 4.5% of the land in the lower forty-eight states.[44] While much of this area contains trees (some 75 billion trees cover approximately one-third of metropolitan areas), this "urban sprawl" is severely fragmenting surrounding forests as urbanization chops up natural areas at its periphery.

Although certainly younger and more fragmented on average than before the European onslaught, forests have managed to survive the worst ravages of the over 315 million people who now call the United States home. That "survival" contrasts starkly, however, with the death throes of its native prairies. At the time of European settlement, it was the largest vegetation province in North America, covering some 160 million hectares (395 million acres). Estimates of the proportion of native prairies destroyed since then are up to 99.9%,[45] mainly due to overgrazing and cropping. This amounts to about 1.7 million square kilometers (660,000 square miles) over all of North America. Tallgrass prairie, occupying much of the central United States east of the Great Plains, has suffered the worst beating, although mixed-grass prairie, to its west where it blends with shortgrass prairie, is not far beyond. In Nebraska, over three-quarters of it is gone. Shortgrass prairie, on the poorer soils of the western Great Plains, has not fared much better, although remaining pockets are the only prairie type existing largely on public land. Even the surviving remnants are having a rough time—grazing and "recreation" are seriously threatening the last bits. Restoration is typically not an option, because once destroyed, prairie soils can take centuries to recover.

Of course, the implication is that much of the dependent flora and fauna is threatened with extinction. Over 700 grassland species are at risk of extinction, and prairie-dependent birds have declined by as much as 65%. The once-widespread biomass phenomenon that was bison herds is no more (although pockets persist and are growing), and the Audubon bighorn sheep and prairie grizzly bear are long gone.

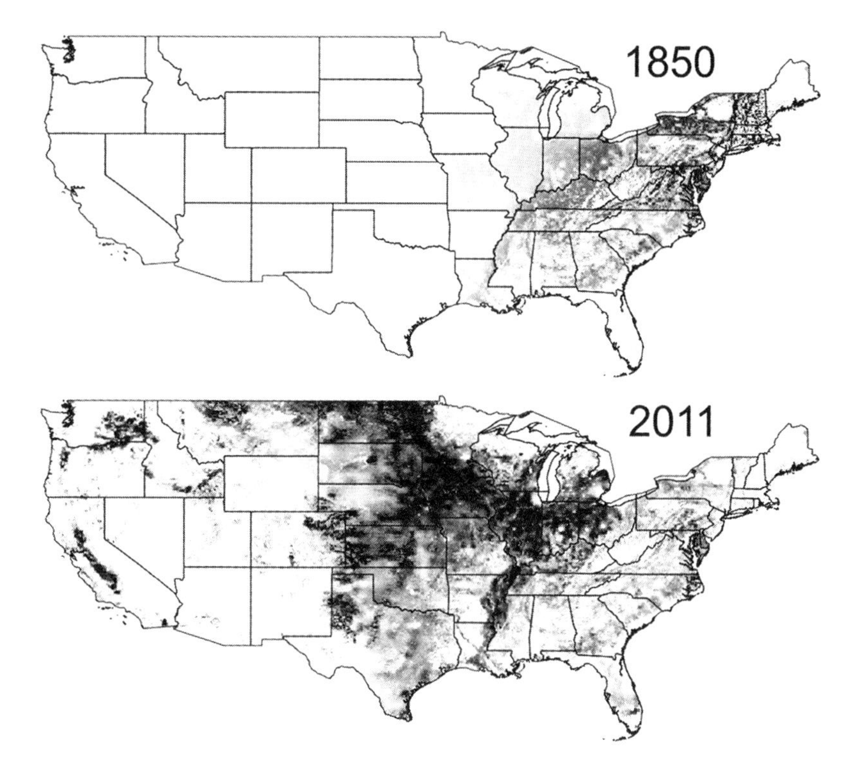

Change in the distribution of agriculture in the coterminous United States from 1850 to 2011. Pixel intensity (white to black) represents the proportion of land area dedicated to cropping. Data updated from N. Ramankutty and J. A. Foley, "Estimating Historical Changes in Land Cover: North American Croplands from 1950 to 1992," *Global Ecology and Biogeography* 8 (1999): 381–96; provided courtesy of Dany Plouffe and Navin Ramankutty.

Let's not forget those once-superabundant prairie dogs and the black-footed ferrets that depended on them, swift foxes, and ferruginous hawks, although our children probably will because it is likely that many of these species will either be extremely rare or even gone by the time they grow up.

Today endangered species are not just restricted to America's hyper-transformed prairies. Overall, about one-third of all US native land plants and animals are at risk of extinction,[46] although there are insufficient data to assess whether this has been getting worse over the last fifty or so years (we suspect it has). According to the US Fish and Wildlife Endangered Species Program, there are 1,400 listed (Endangered and Threatened) plant and animal species, for which most have special protection under the Endangered Species Act (ESA). The problem is at least for some groups that the ESA does not adequately cover the full breadth of endangered species recognized by the international "gold standard" of species status—the Red List of Threatened Species.[47] For example, over 40% of Red Listed US birds are not listed by the ESA.[48] But the discrepancy is not restricted to birds; in total, there are a hell of a lot more US Red Listed species than are listed by the ESA—some 64,000 species, of which more than 20,000 are threatened with extinction. Worse yet, as in Australia, the focus tends to be on the extinction of species, when the extermination of populations is at the moment much more extreme and serious. Population extinction is virtually always a prelude to species extinction, and it is populations that provide humanity with its essential income from biowealth—such as climate regulation, control of crop pests, provision of pollination to agriculture, recycling of nutrients, and support of oceanic fisheries production. In P.R.E.'s fieldwork in the United States, he has repeatedly watched populations of butterflies disappear but has never personally witnessed the loss of a species.

A little thought experiment will make this clear. Suppose every species in the United States, Australia, and the rest of the world was suddenly reduced to a single viable* population that was permanently protected. The problem of species extinction would be permanently solved—no more species would go extinct. So would the problem of *Homo sapiens* be solved, since once stored food was exhausted, all human populations would perish.

Neither are America's freshwaters doing well. Admittedly, there have been improvements since the infamy of rivers catching fire and the legacy of poisoned waterways and groundwater often left by "wholesome" US companies like General Electric.[49] Today in the United States, some 60% of farmland streams and 80% of urban and suburban streams have at least one contaminant at concentrations that threaten aquatic life. A recent study demonstrates that since 1898, North America has lost at least thirty-nine freshwater fish species, which is a rate about 900 times faster than expected based on the fossil record.[50] One of the most remarkable threats to watercourses in the United States

*Guaranteed to persist for a long time into the future.

Top: Among the best-documented population extinctions were those of three populations of the checkerspot butterfly *Euphydryas editha* that P.R.E studied for decades at Jasper Ridge on the campus of Stanford University. Climate change in the San Francisco Bay area of California was a major factor in their disappearance. *Bottom:* Code numbering of a checkerspot butterfly in a capture-mark-release-recapture experiment that allows population size to be estimated. Field biologists often have dirt under their fingernails. Photos by Paul R. Ehrlich.

are concentrated animal feeding operations (CAFOs). For example, thousands of swine are often confined in a small area and their feces collected in big ponds. Fermentation can lead to geysers in the mess, and occasionally retaining dams will break. In 1995 a CAFO caused what was up until then the largest environmental spill in the history of the nation—more than twice that of the disastrous Exxon *Valdez* oil spill. A North Carolina hog manure lagoon ruptured and released nearly 100 million liters (26 million gallons) of effluvium into the New River. It killed every living creature in the river, including fish by the millions. The feces also breed microbes that produce neurotoxins that can disorient people exposed to the floods of pig feces—a phenomenon known technically as the "*Fox News* Effect." While CAFOs do not contribute much to global problems, they are quite capable of making local areas inhospitable to human beings.

Also in the United States, about one-third of groundwater wells and 20% of streams have one or more contaminants that are considered unsafe for human health.[51] As we write this, much of America's West is suffering from a record drought, likely worsened by climate disruption, and Americans are now draining their precious groundwater resources. Even in the great aquifer system of arguably America's most productive region—the Central Valley of California—withdrawal is exceeding renewal,[52] thus threatening the long-term future of that productive food bowl, which is also threatened by global warming reducing the snowpack in the Sierra Nevada, whose melting provides critical water to California's agriculture. Expansive urbanization in the Central Valley paving over all that fertile soil is not helping matters. And over much of North America, there is growing concern about the possible impacts of "fracking" on groundwater quality—although comprehensive answers are not yet in.[53] What seems more certain is that the vast natural gas boom connected with fracking is not going to exploit gas as a "bridge" fuel, but instead will likely just exacerbate the flux of the greenhouse gases like methane, and eventually CO_2, into the atmosphere—or as Naomi Oreskes calls it, "a green bridge to hell."[54]

The ecological damage in the United States is not restricted to the land and freshwaters, of course. A long history of marine overexploitation in North America is still evident today. In addition to the ecological cod travesty of the Grand and Georges Banks off North America's eastern coast, the Gulf of Mexico is now largely overfished. Massive declines in sharks over the last fifty years there have completely upset the delicate ecosystem balance of that once-productive region. Indeed, another sort of "mesopredator" release has already occurred—the huge decline in large sharks has stimulated an explosion of cownose rays that have gone on to destroy scallop fishing in the southeastern United States.[55] Other examples of marine ecosystem disruptions are plentiful. The eradication of sea otters from much of the western coastline caused explosions in sea urchins, which went on to eat away (quite literally) the vast kelp forests that supported much of the coastal ecosystem. It is important to consider the implications of this loss. Kelp forests are some of the most important features of coastal ecosystems, for just as terrestrial forests support millions of other species, so too do their marine versions support countless others, many of which supply humans with food. While recoveries have occurred, kelp forests now occupy only a fraction of their original range. Most recently, warming waters off the coast of the Pacific Northwest have driven salmon farther north into Canadian coastal waters, depriving local fishers of their livelihoods.[56] Additionally, overfishing in estuaries, bottom-trawling, and offshore fishing has essentially pilfered the majority of the marine bounty that made North America prosper in her early days.

While US air quality has improved markedly over the last forty years with the increase in pollution regulations on transportation vehicles and industry, the world's latest atmospheric demons—greenhouse gases—are mostly getting worse. As the prime determinant of climate disruption, these gases (including carbon dioxide, methane, and nitrous oxide) are pushing the planet toward a dangerous climatic tipping point. We are already seeing the results: rising sea levels are threatening low-lying countries in the southern Pacific and Indian Oceans, climate extremes such as heatwaves[57] and droughts are increasing in severity and frequency, ocean acidity is rising (threatening the very base of the marine food web), and, of course, the entire world is heating up (and faster in the Northern than Southern Hemisphere). But this book is not about the evidence for, the arguments surrounding, or the politics of climate change—there are plenty of excellent treatises of this topic that we recommend.[58] Do not be sucked in by the denialist machine,[59] however. Human climate disruption is not a belief, it is not a conspiracy, and it is *definitely* something about which you should be terrified (remember Sandy, the amicably named hurricane of 2012?). That means being terrified into taking action—writing letters to newspapers, not voting for any denier politicians, donating to environmental nongovernment organizations, or whatever fits your proclivities.

That over 97% of scientists worldwide (that's tens of thousands of them) concur that climate disruption is a result of human endeavor[60] suggests, at least to some ideologically right-wing people, that there is a massive, worldwide, left-wing conspiracy to dupe the masses. While this tired accusation is most commonly made about climate scientists, it applies across nearly every facet of the environmental sciences whenever someone disagrees with what one of us says. First, it is essential to recognize that most scientists are just not that organized. While neither of us has yet forgotten to wear trousers to work, most of us are still far away from anything that could be described as "efficient" and "organized." Such is the life of the academic.

More importantly, the idea that a conspiracy could form among scientists ignores one of the most fundamental components of scientific progress—dissension; and can we dissent! The scientific approach is one where successive lines of evidence testing hypotheses are eventually amassed into a concept, then perhaps a rule of thumb. If the rule of thumb stands against the scrutiny of countless tests, then it might eventually become a theory. A theory is not, as many believe, merely an untested model of how something works—it is instead a massive body of tested evidence. Some theories even make it to become the hallowed law, but that is very rare indeed. In the environmental sciences, one could argue that there is no such thing as a law. Well-informed non-scientists might understand, or at least appreciate, that process, but few people outside the sciences have even the remotest clue about what a real pack of bastards we can be to one another. Use any metaphor you want—it applies: dog-eat-dog, survival of the fittest, jugular-slicing ninjas, or brain-eating zombies in lab coats.

The first tunnel of pain is in the review process itself. Ask any PhD student after receiving the referees' (peer reviewers') comments on his or her first paper. Most often it involves an outright rejection, typically accompanied by some caring and supportive words like "fail," "flawed," and "nonsense." It does not improve either as the scientist progresses through her or his career—we just become numb to the pain and soldier on. In other words, if there are any chinks in the armor of the evidence for any particular phenomenon, other scientists are the first to expose and exploit them. In fact, many scientists have built their entire careers out of destroying the work of others. Scientific conspiracies are therefore scientifically implausible—and, of course, *any* conspiracy that depends on tens of thousands of people keeping a secret is even more implausible.

Back to greenhouse gases. Globally, emissions are continuing to rise at a frightening pace, and the latest (2013) Intergovernmental Panel on Climate Change[61] report demonstrates that we are even beyond the worst-case scenarios it predicted in 2007. The United States and the European Union have recently made small reductions in total emissions of about 1.7% and 1.9% in 2012, respectively.[62] The reasons for these declines include rising use of natural gas over coal to produce electricity, a reduction in oil consumption, and a warm Northern Hemisphere winter. This is, of course, some mildly good news—but let's take a look at the bigger picture. In 2009 the United States was the second-highest emitter of carbon dioxide emissions at over 5 gigatons (China was the clear leader with nearly 7 gigatons). Put another way, this was 18% of the *world's total emissions*, despite the United States making up only 4.4% of the world's population! In addition to being one of the largest emitters, this makes the United States also one of the world's largest per capita emitters. Clearly something is amiss here. These emissions are driven to a large degree by the size of the US economy (currently the largest in the world), but the per person wastage is, quite frankly, disgusting.

And what about little old Australia? Australia's economy is only about 10% the size of the US economy, and there are fourteen times as many people in the United States than in Australia. Unfortunately it turns out that Australians are emissions pigs too. In 2006 Australia had the highest per capita emissions among the Organisation for Economic Co-operation and Development (OECD) countries, being surpassed only by five non-OECD countries (Bahrain, Bolivia, Brunei, Kuwait, and Qatar). Indeed, that year, Australia's emissions per person per year were around twice the OECD average, and more than four times the world average.[63] Australia also exports hundreds of millions of tonnes of coal overseas each year*—much of it to furnish China's expanding

*274 million tonnes in 2011: www.dfat.gov.au.

economy. To get an idea of how much coal is involved, Australia exports each year the equivalent of a one-square-meter pile that would stretch 8.2 times around the Earth at the equator.* This translates into about 836 million tonnes of carbon dioxide equivalents released in one year[64]—which is more than the entire annual production of greenhouse gas emissions from Germany. Apart from the coal barons raking in the resulting profits, every Australian should be ashamed and disgusted by this wanton pollution.

Despite being one of the driest continents on Earth, Australia also has a huge water footprint.† For internal domestic use (i.e., not including agricultural and industrial uses, or water imported directly or within other goods), Australians use about 341,000 liters (90,000 gallons) per person per year,‡ which is six times the global average (57,000 liters, or 15,000 gallons per person per year).[65] In the United States, the internal domestic use is a little better at 217,000 liters (57,000 gallons) per person per year, but the United States' massive industrial and agricultural sectors push that per capita water use up to nearly 2.5 million liters (660,000 gallons) per year. Based on data from 1997 to 2001, the United States therefore had the highest per capita water consumption among twenty-one developed and developing nations comprising most (6 billion people) of the world's population.[66]

Agricultural production is one of the chief consumers of freshwater around the world. For example, the global average virtual water content of rice (paddy) is 2.29 million liters per tonne (605,000 gallons) produced, and for wheat it is 1.33 million liters per tonne (350,000 gallons). Growing crops for biofuel in particular has a huge water footprint—depending on the crop in question, it takes an average of 1,400–20,000 liters (370–5,300 gallons) of water to produce just one liter of biofuel![67] If an agricultural product comes from livestock—say, meat, leather, or wool—the water content is typically much higher because of the feed required to keep the animal alive. For example, it takes about three years to raise beef cattle to slaughtering age, with an average of 200 kilograms (440 pounds) of boneless beef produced per animal. This requires about 1,300 kilograms (2,900 pounds) of grains,

*Based on a bulk bituminous coal density of 833 kg/m^3.

†Defined as "the total volume of freshwater that is used to produce the goods and services consumed by the people of the nation."

‡Calculated from 1997–2001.

7,200 kilograms (16,000 pounds) of pasture or hay, and 31,000 liters (8,200 gallons) of water for drinking and cleaning. This means that the total amount of water required to produce 1 kilogram (2.2 pounds) of beef is about 15,340 liters (4,050 gallons)![68] For Australia and the United States, which collectively have about 120 million or so cattle at any one moment, the water footprint alone should at least be cause for concern the next time you tuck into a steak dinner.

Export-focused agriculture has other big problems in addition to the inefficient consumption and export of precious water resources. For example, soil degradation from erosion, depletion, chemical fertilization, salination, and toxification is a massive global problem—approximately 12% of global agricultural supply has declined from soil degradation alone over the past fifty years.[69] Another sobering reminder of agricultural inefficiency is the situation in aquaculture. The overall global average conversion ratio of fishmeal (commercial product made from grinding whole "trash" fish and other seafood, trimmings, bones, etc., derived primarily from small, bony, oily fishes) in aquaculture is about 0.70, which means that it takes about 1.43 kilograms (3.2 pounds) of fish-based meal to produce 1 kilogram (2.2 pounds) of farmed fish[70] (although this is still more efficient than livestock production). For places like South Australia with a thriving overseas market for farmed bluefin tuna, a good part of the economy is therefore based on sending its marine resources overseas. In Australia, where resource extraction and export—coal, minerals, seafood, topsoil, and water in livestock—are responsible for most of the country's economic success, the future prospects of maintaining the current standard of living is highly threatened as these resources become depleted. Inefficient use and exportation of Australia's rare water resource alone should ring alarm bells. In a political climate where science, technology, and manufacturing are of low priority, the writing is on the wall if this "mining" mind-set continues. The United States has the clear advantage over Australia here, since it does not have the same urge to become a less-developed country, primarily exporting natural resources from declining stocks, although its recent and insane fossil-fuel boom could indicate a turn in that direction.

6 Sick Planet, Sick People

No amount of medical technology will enable us to have healthy humans on a sick planet.

THOMAS BERRY, quoted in *Radical Homemakers* by S. Hayes (2010)[1]

Let's face a cold, hard fact. The concept of the essential role of functioning ecosystems for human survival is completely foreign to most people,[2] and any demonstration of ecological damage is so terminally boring that at best we can expect a mild "tut, tut" from even the most well-informed. Modern society is utterly unconcerned with our plight, such that environmental scientists are constantly branded as doomsday propagators or as out-of-touch prophets of a future that might at worst be only slightly more uncomfortable than the technological comforts that modern life now affords. The second cold, hard fact is, however, that we are unfortunately not nearly as frightened as we should be. If, for example, everyone understood just how much Earth's overall climate has shifted in the past, from the heating that ended the last glacial stage to a likely total freeze into a "snowball Earth" period much earlier, humanity collectively might not sleep so soundly. In fact, virtually everyone would be awake to civilization's peril and would, for instance, be demanding an all-out effort to reduce population size humanely, help the poor while restricting consumption among the rich, rapidly move to transition from fossil fuels to more environment-friendly energy technologies, and take the other steps that might give our grandchildren a chance.

Climate disruption is, as we discussed, one of the greatest monsters of our time, and we are not talking about the imaginary supernatural monsters of the pious. But this is not a book about climate change per se. We have outlined the environmental histories of our two countries since before the onset of the Holocene era and described in detail just how much our biowealth has been compromised during the Anthropocene[3]—the Age of Human impact, since about the last 10,000 years. Now we need to add another layer of ecological destruction to the tableau of degradation we have begun to paint. We do this not to darken further our already penumbral reputations; rather, it seems to be common element of human nature that positive action is difficult unless one can demonstrate a direct threat to individual health and wealth. So here comes the health part of the story.

It has taken most of humanity a while to realize that what we do to the planet, we eventually end up doing to ourselves. This concept should not be difficult to grasp—not many people would normally go to the toilet in their kitchen; but at the global scale, the link between toxic behavior and human health is almost completely ignored by the masses. As it turns out, there are quite a few examples of how we are rapidly making ourselves more susceptible to killer infectious diseases and toxins simply by our modification of the planet. If we look at the most obvious problems caused by the residue of development—air and water pollution, as well as soil and food contamination—it probably will not be much of a surprise that pollution causes about one-quarter of the global "burden of disease,"[4] mainly among the world's 2 to 3 billion poor.[5] For instance, today almost 1 billion people currently lack safe drinking water, and some 2.5 billion people do not have adequate sanitation.[6] But it most likely *is* a surprise to learn that environmental health hazards such as climate change,[7] freshwater shortages,[8] and fisheries degradation[9] compromise human health *to an equal or greater extent* than pollution per se.[10]

There are many other subtler and unusually complex ways in which human health is eroded by environmental damage. For example, deforestation can increase mosquito-biting rates and raise the incidence of malaria.[11] The rise of diseases like West Nile fever in the United States is another example of how human encroachment on nature exposes us to novel pathogens. Deforestation also increases the risk of catastrophic floods, which can kill hundreds of thousands per decade, displace millions more, and cause trillions of dollars in damage.[12] Floods, like the one that covered nearly a fifth of Pakistan in 2010, can lead to vast outbreaks of disease like schistosomiasis, malaria, leptospirosis, dysentery, cholera, hepatitis, and typhus within the displaced populations. This occurs primarily because of the resultant high-density living in temporary refugee camps with inadequate sanitation and health services, and because of temporary increases in the habitat availability for the pathogen's hosts (e.g., insects and snails).[13] Further, because most emerging infectious diseases are caused by zoonotic (animal-borne) pathogens, of which more than 70% are known to derive from wildlife,[14] increasing habitat fragmentation and penetration by expanding human populations into formerly isolated regions risks higher probabilities of novel diseases transferring from their animal reservoirs to people. Of course, modern transport systems speed the spread of plagues.[15] In the Middle Ages, if a ship carrying individuals with smallpox left India for Japan, everyone on board was dead or immune by the time it arrived. A Japanese smallpox epidemic did not result. Contrast this with the story of a handsome blond flight attendant on Air Canada, "patient zero," who almost single-handedly spread HIV/AIDS around the world. In an age where millions of people circulate globally at 600 miles per hour, the old anti-epidemic tool of quarantine seems ever less effective.

The health threat does not end with diseases—it extends to a form of pollution that so far has escaped the attention of the wider public. There are frightening symptoms traceable to the release of hormone-disrupting toxic substances, increasing reasons for concern as more is learned about the ways in which early exposure to bioactive chemicals can influence development and survival. Besides assaulting many species that make up our life-support systems, toxics could be shifting the human sex ratio, causing developmental problems in children, and possibly reducing sperm counts. Fears about toxins in food, although still a great concern in rich nations, have in China returned to the 1930s of Kallett and Schlink's *100,000,000 Guinea Pigs*.[16] As one Chinese blogger recently said: "Every day I ask myself, what is safe to eat? The pork is laced with clenbuterol [a steroid used to bulk up livestock], the beef and lamb contain other toxic additives; and we don't drink the milk."[17] Having both regularly visited China, we can say with some authority that it is best not to think about how the food is produced lest you go hungry for the duration of your stay there. But these trends remain mostly unrecognized. Even in Australia, where toxification is unheard-of by most people, there are some pretty filthy sites, such as Homebush Bay[18] (the site of the 2000 Sydney Olympics), Sydney Harbour[19] (Port Jackson), the Derwent Estuary[20] in Hobart, Tasmania, and the infamous, toxic moonscape of Queenstown[21] in western Tasmania.

Rarely mentioned in the media or by politicians, the appreciation of these problems is rudimentary. This is especially the case as recent research has shown that a basic rule of toxicology—"the dose makes the poison"—is often dead wrong (pun intended). Yes, it is true that even table salt can kill you if you eat too much of it. But there are other novel chemicals, those "endocrine disrupting compounds" (EDCs) just mentioned, that can do nasty things to you at high doses but can have different negative effects at very low doses. These low-dose adverse effects can increase the probabilities of altered sex determination and other developmental defects, behavioral changes, developing cancers, and more. If one plots the impacts of many toxic chemicals against the dose absorbed, there is a continuous increase of danger as the dose goes up. Such a dose-response curve is said to be "monotonic." But for chemicals that behave like hormones—EDCs—the response might at first increase and then decrease—the function can change sign from positive to negative, or negative to positive. This happens for several reasons, most commonly because hormones and hormone-like chemicals alter gene expression at very low doses; but at high doses, these receptor-mediated responses are either saturated or shut down. This often produces U- or upside-down U-shaped curves, which in the jargon of biology are "non-monotonic dose-response curves." Such nonlinearities—that is, situations where the relationship between cause and effect are not constant and the output of a system does not change in direct proportion to the input—are characteristic of many environmental situations: for example, the hypothesized nonlinearity in the relationship between economic development and environmental damage known as the Kuznets curve, which we will visit in later chapters. In toxicology, they are too frequently ignored. EDCs have come into public-health prominence largely due to a book published in 1996, *Our Stolen Future* by Theo Colborn, Dianne Dumanoski, and J. P. Myers.[22] In the tradition of *Silent Spring*, they sounded the alarm, were attacked by the chemical industry, and were subsequently proven correct—as a recent statement by the Endocrine Society demonstrated.[23]

A huge mass of evidence is building that should alert humanity to the risk potential of toxifying Earth with synthetic chemicals, especially EDCs. As Linda Birnbaum, director of the US National Institute of Environmental Health and the National Toxicology Program, said in 2012:

> The question is no longer whether non-monotonic dose responses are "real" and occur frequently enough to be a concern; clearly these are common phenomena with well-understood mechanisms. Instead, the question is which dose-response shapes should be expected for specific environmental chemicals and under what specific circumstances.

The impacts of low doses of EDCs can be graphic. For example, take the case of laboratory mice exposed to low doses (1 μg/kg/day) of on artificial estrogen (female hormone) diethylstilbestrol, once used to treat pregnant women in the false hope it would prevent complications. It is now known to cause an otherwise rare vaginal cancer in the daughters of the women treated, as well as other nasty effects on both daughters and sons. At first, the low-exposure mice had lower body weights compared to control mice. But after two months, exposed mice surpassed the controls and became unusually obese.[24]

Other examples abound. Infant mosquitofish subjected to small doses of 4-nonylphenol, an industrial chemical used in large quantities in various operations, produced adults that all had female characteristics, although the normal 50-50 male-female ratio persisted in the sex organs.[25] In the wild, a high proportion of alligators in a Florida lake polluted with EDCs had developmental abnormalities leading to sterility.[26] Even more worrying is anecdotal evidence of possible human consequences of exposure to EDCs, apparently influencing the interactions of hormones involved in sex determination. For instance, in one Canadian village adjacent to a large complex of petrochemical plants, the ratio of boys to girls at birth has recently declined from the normal of about one-half to one-third—although EDCs have not been conclusively demonstrated to play a role in that decline.[27] In some parts of the Arctic, a notorious sink for EDC pollutants including PCBs,* DDT, and flame-retardants, twice as many girls are being born as boys.[28] There apparently has been a general decline in the proportion of boys being born in Japan and among US whites (but not Af-

*Polychlorinated biphenyls.

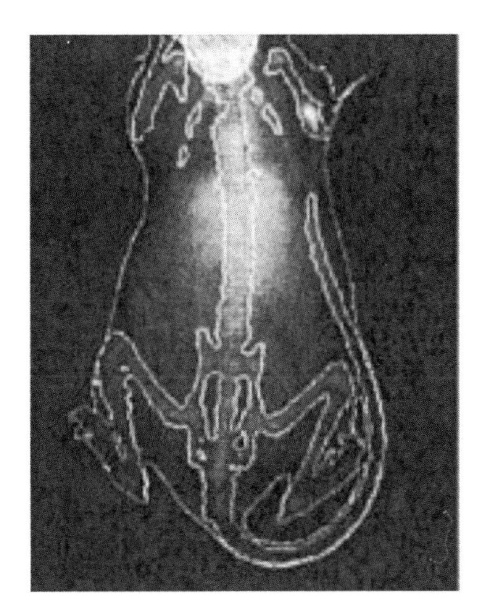

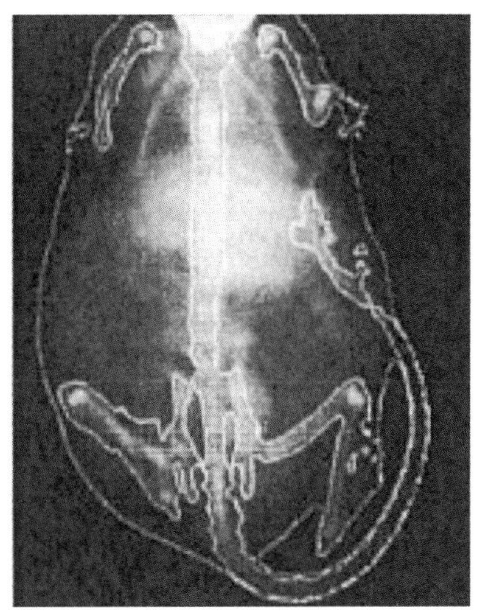

Endocrine disruption: exposure to diethylstilbestrol (0.001 mg/kg/day) (*right*) causes an increase in percentage body fat in young mice relative to controls (*left*). From R. R. Newbold, E. Padilla-Banks, R. J. Snyder, and W. N. Jefferson, "Developmental Exposure to Estrogenic Compounds and Obesity," *Birth Defects Research (Part A)* 73 (2005): 478–80. Reprinted with permission from John Wiley & Sons Ltd.

rican Americans)—due to more male fetal deaths.[29] Assigning blame to one or more chemical compounds or other factors in such cases is extremely difficult, since many of them might interact to produce damage that is more than the sum of their individual effects. Such synergisms often defy identification—even highly suspect ones where exposure data are relatively abundant, as between smoking and asbestos, have required multiple analyses. A recent thorough review of the scientific literature documented the pervasive evidence for low-dose effects of EDCs in populations of human beings and of wildlife.[30] One of its conclusions (conservative in our view) is that "regulatory action to minimize or eliminate human exposures to EDCs, could significantly benefit human health."

Today humanity is faced with a series of epidemics in which toxics could be involved: asthma, autism, hormone-related cancers, attention deficit/hyperactivity disorder (ADHD), heart disease, autoimmune diseases, obesity, type II diabetes, and learning disabilities, just to name the most common. Of course, toxins might be found innocent in many cases, but is it wise to keep releasing them *ad lib* into the envi-

First public test of insecticide machine: beachgoers are sprayed with DDT as a new machine for distributing the insecticide is tested for the first time. Jones Beach State Park, Wantagh, Long Island, New York, USA (1945). Photo credit: © Bettmann/CORBIS.

ronment on the assumption they are not involved? Eight decades after *100,000,000 Guinea Pigs*,[31] the behavior of the food and chemical industries, Big Pharma, and the tobacco industry seems not to have changed all that much, considering the bisphenol-A (BPA) disaster,[32] the many hundreds of thousands of annual deaths from cigarette smoking in the United States alone, the type 2 diabetes epidemic,[33] and everyone's general level of exposure to EDCs. Beyond this, there are the horrific potential problems associated with growing, complex pollution from destruction of electronic wastes (from billions of discarded mobile phones, computers, etc.) and burgeoning nano-products that easily penetrate our cells.[34] If you want to be scared out of your wits by these threats, just refer to Julian Cribb's brilliant book, *Poisoned Planet*.[35]

If climate disruption starts to run away from our ability to mitigate and adapt to it, there are at least proposed "solutions" in the area of geoengineering. They are likely to be extremely expensive and politically and environmentally hazardous in the extreme[36]—but at least we might be able to do something. If it turns out that the diseases some believe are now connected to toxification continue to increase, or, say, if new and more virulent forms of cancer pop up and become ubiquitous, what could be done? There is no backup of an unpolluted Earth to which we can escape. Rachel Carson was properly concerned about the problems of unknown interactions among the mixtures of novel chemicals in the environment. Today, with the potential for millions of possible synergisms among the tens of thousands of compounds already released, even identifying the culprits could be an enormous challenge. If they were identified, removing them from the environment could be an even more monumental task, and most likely prove impossible.

Carson was much concerned with ethical issues related to the insertion of toxic chemicals or other pollutants into other people's environments. Her concerns were generated during her adolescence when two coal-fired power plants turned her hometown into a polluted wasteland. That, of course, is still a concern, but people in general share the responsibility for their desire to have many of the products, including electricity, that result in toxification. In turn, many of the toxins released could be retained with more careful handling and product disposal, but often at higher prices.

If you still think the days of naive DDT-like chemical profligacy are a thing of the distant past, think again. The World Health Organization itself estimates that at least one in every twelve deaths worldwide arises from exposure to toxic chemicals. This means it kills more people each year than do car crashes, malaria, and HIV/AIDS.[37] In *Poisoned Planet*,[38] Julian Cribb argues—with convincing evidence—that the entire planet is affected by toxic pollutants. From the high latitudes of the Arctic and Antarctic, to the bottom of the seas, nowhere on Earth, and no one on Earth, is immune from the toxic influence of human industry. Despite the rise of many legal checks on the production, storage, distribution, and waste management of toxic chemicals since the 1970s in most developed nations at least, the production of synthetic chemicals is today some twenty times higher than it was during the 1960s. There have been over 140,000 new industrial and agricultural chemicals synthesized over the last fifty years.[39] Cribb shows that many of the ecological and health risks of these toxins has been vastly underestimated,[40] partially because the issue is no longer occupying the main environmental radar for most people because we assume wrongly that our existing checks and balances are sufficient.

In our view, there are two basic solutions to situations where the activities of a segment of the human population, often an industry, present a clear and present threat to the human future. A classic example of such a hazard resulted from the manufacture of chlorofluorocarbons. These synthetic chemicals threatened the vital stratospheric ozone layer, whose depletion could have ended all human life. Today the fossil-fuel and synthetic chemical industries clearly could destroy civilization as we know it. One way to prevent this would be society to find ways to alter the behavior of big industry to minimize the impacts on those involved. Society needs energy, but it can no longer afford to mobilize it by burning fossil fuels en masse for most of its needs. John Browne had the right idea when he changed British Petroleum into BP and gave it the slogan "Beyond Petroleum." Society needs some synthetic chemical materials, but it can no longer afford to manufacture and distribute them as incautiously as it now does.

Browne's approach did not last at BP, but perhaps converting the producers of toxins to "green chemistry" (design of products and processes that reduce or eliminate the use and generation of hazardous substances) could be a step toward a permanent answer. The idea would be to make the basic ingredients of the American or Australian version of prosperity less toxic or even nontoxic. This would shift the environmental issues in molecular design up front, considering not only the chemicals to be purposely created, but also any intermediate compounds produced or employed in the process. Green chemistry has great promise as a path toward inherently safer materials, but it faces serious impediments. One of the biggest and most immediate is that the field currently lacks the capacity, in terms of trained people, to ask the right questions and generate workable solutions. At a time when US and Australian science education is on the wane and when industrial jobs in both countries are being sent overseas in search of cheaper labor and higher profits for the wealthy (or, mainly in the case of Australia, failed to develop in the first place), this kind of scientific industrial revolution could breathe new life into a sector of industry that might otherwise be doomed to obsolescence and decay.

In an era of exploding human populations[41] and escalating resource scarcities, the pressure to produce many substances with known or possible toxicities (e.g., for pest control) will probably grow, and grow much more rapidly than toxicologists, developmental biologists, and ecologists can even estimate the risks for. This is especially so in view of the myriad of potential but unknown synergies possible in the now-global stew of toxic substances. One step that could help would be to require tight benefit-cost analyses on any chemical before it goes into production. But this would likely be expensive and difficult to monitor—and the problems of assessing potential benefits could easily prove insurmountable. Other steps involve invoking the precautionary principle beyond moving toward green chemistry, by prohibiting all discharge of persistent organic pollutants, by shifting the burden of proof from having to demonstrate that a chemical is dangerous to requiring it be shown to be safe (an almost always difficult task), and rapidly phasing out production of the most dangerous classes of compounds, such as organochlorines.

A more fundamental answer, we believe, is that the global system needs to be rescaled. It is now in serious overshoot, with more than one Earth now required to support just today's population indefinitely at its current rate of resource consumption. Humanity seems to be moving fast away from sustainability and showing diminishing marginal returns to complexity that Joseph Tainter suggested[42] are some of the main warning signs of societal collapse. Such diminishing returns are now evident everywhere, affecting virtually all the resources that civilization needs to persist.[43] A gradual humane reduction of human population size[44] and a reduction in the overconsumption of the rich would greatly alleviate the pressure on the chemical industry to keep churning out dangerous chemicals. We imagine that Rachel Carson, with her population-oriented discussion of pesticide problems—pointing out that "populations are kept in check by something the ecologists call the resistance of the environment, and this has been so since the first life was created"—would be very sympathetic to such a rescaling.

Of course, as Carson's critics realize, adequate measures to confront a growing toxic threat would have serious financial consequences for the chemical industry and could result in considerable inconvenience for consumers. To avoid that, one possibility is to pursue business as usual and count on luck to save civilization. Maybe no truly lethal synergies will turn up or no new chemical will be released and become global before it is discovered to cause an untreatable cancer. It could be that the decline in male births will halt or reverse before causing serious social problems. Maybe the poisoning of nonhuman organisms will not cause sufficient collapse in ecological systems to destroy vital ecosystem services—as appears to be happening with the pollination service of bees today[45]—and bring down civilization. Perhaps advances in molecular biology will develop methods to identify and neutralize dangerous novel compounds or to cure any serious diseases that appear. Perhaps they will not. Is it wise not to take substantial defensive measures? In democracies, the decision should rest ultimately with the citizens; we think it is crystal clear what Carson would have recommended.

But her recommendations would doubtless seem almost far-fetched in the face of continued besmirching of the environmental sciences by the corporate forces of evil profiteering from environmental degradation. Yes, it is happening even today. Tyrone Hayes*of the University of California, Berkeley, has suffered Carson-like abuse by flacks for the chemical industry for demonstrating nasty low-dose impacts of the near-ubiquitous herbicide Atrazine in both the laboratory and in nature. Atrazine is manufactured by Syngenta,[46] which has bankrolled a multimillion-dollar campaign, hiring pundits to lie about Atrazine and to defame Hayes. Unlike in Europe, where it has been banned since 2004, Atrazine is still legal in the United States and Australia, despite the mounting evidence for its detrimental health and environmental impacts. It is insane that we are still using it given the well-established history of once supposedly "miracle" chemicals like DDT. The campaign resembles that of the well-funded fossil-fuel industry assault on massive evidence of climate disruption and also the tobacco industry's long-running storm of lies about the safety of smoking.[47] Indeed, it features some of the very same characters, such as one infamous *Fox News* hack[48] and paid advocate of Philip Morris tobacco, ExxonMobil, and other corporate monsters. Many of the attacks on Carson focused on her gender;[49] those on Hayes sometimes related to his skin color.[50] He has been, as was Carson, scientifically vindicated and, like Carson, is one of our heroes.

*Hayes gives a great lecture—if you ever get the chance to see him speak in person, take it. You will not be disappointed.

7 The Bomb Is Still Ticking

Democracy cannot survive overpopulation. Human dignity cannot survive it. Convenience and decency cannot survive it. As you put more and more people into the world, the value of life not only declines, it disappears. It doesn't matter if someone dies. The more people there are, the less one individual matters.

ISAAC ASIMOV on Bill Moyers's *World of Ideas*, Public Affairs Television, New York (1988)

Thus far we have shown you what our now very large human population has done to the United States and Australia, and to the planet in general. It might come as something of a surprise to most that of all the human beings who have ever lived since our modern species first evolved some 200,000 years ago, around 14% of them are still alive today![1] This rather surprising fact demonstrates just how much the human population has increased in the last century alone. It therefore bears repeating—and repeat we shall *ad nauseam*—that there is a serious assault on science and humanity beyond the physical damage of our ecosystems: that is, the systematic claim that population growth is either beneficial, or at least is not seriously harmful. Obviously, expanding human numbers, and our already huge overabundance, is itself a main cause of the "perfect storm" of environmental disasters civilization now faces. There is a major difference between the two assaults, however, in that many of those who think the population can and should grow forever are not united by greed or even ideology, but by a lack of understanding of basic science. Roman Catholic bishops fight contraception* and abortion to protect their ideological base. In

*Even though a recent reconsideration by Pope Francis of the role of condoms in reducing disease risk is encouraging but entirely inadequate.

so doing, their main damage in the past has been to cripple US government efforts (and those of others) to spread family planning around the world by misleading and intimidating politicians of other persuasions. Today the Catholic Church is a major henchman in the US Republican war on women, especially in working to impede access to safe abortion. In Australia family planning occurs, but it is still far too difficult for a woman to obtain an abortion should she choose to do so. The actions of politicians in both countries have tragically condemned millions of women to injury and death in unsafe abortions, and helped to perpetuate poverty in developing nations. Based on data from 2008, it has been estimated that there were about 86 million unintended pregnancies globally of the 208 million women who became pregnant that year (or about 41% of all births).[2] Of those 86 million unintended pregnancies, around 11 million were miscarried, 41 million were aborted, and 33 million resulted in unplanned births.[3] Imagine what a different world it would be if those pregnancies could be avoided altogether.

The crimes of these putative emissaries from supernatural beings are unforgiveable. If only the bishops were not totally ignorant about human sexuality and the perfect storm of problems that civilization now faces. If they were ethical men, they would then quickly see through the Church's antiquated and immoral notions and desert the trenches of the war on women. It is noteworthy that Catholic laypeople generally use contraception and abortion at about the same rate as non-Catholics in the same nations. Indeed, mainly Catholic nations in Europe are among those with some of the lowest birth rates on the planet. Of course, many of those unfazed by the population explosion are not Catholic, including multitudes of businessmen and innumerate economists who imagine that ever-increasing numbers of people are necessary for economic prosperity, and that the population (or economy) can grow forever. Greed mixed with profound ignorance is certainly one element underlying much of the ideological perpetual-growth doctrine.

To a large extent, refusal to recognize that continued population growth is a serious threat to the future of civilization can be blamed on the failure of educational systems to bridge key parts of the culture gap[4]—the growing chasm between what we each know as individuals and all of the knowledge society possesses. That gap leaves many "well-educated" people ignorant of today's crucial environmental problems.

What do people need to know to build the necessary bridges over the principal components of the gap? First, they must understand that population growth is one of three major drivers of the deterioration of human life-support systems. This is hardly rocket science;* the pressures that a population places on the environment are a product of the number of people, multiplied by average per capita consumption, multiplied in turn by how environmentally well or poorly that consumption is serviced.[5] This is the basis of the famous I = PAT equation,[6] where *P*opulation, multiplied by per capita consumption (*A*ffluence), multiplied by the *T*echnological and socioeconomic choices made to service the consumption give a rough estimate of a group's *I*mpact on its life-support systems. Such "density feedback" is in fact a fundamental characteristic of populations of *all* species,[7] not just humans—it is a truism that as population density increases (number of individuals per unit area), per capita resource availability declines after a certain point. In other words, each individual has, on average, access to a little less food and shelter, notwithstanding the disparity in wealth distribution among the citizenry and the ability of human beings to discover and mobilize new resources. The upshot is that because of these per capita resource declines, aggressive interactions can increase, inequalities become more pronounced, and ultimately battles often ensue. High densities also assist the spread of pathogens. In nonhuman systems, this density feedback eventually leads to a lowered probability of survival because of lack of food and shelter, increased probability of dying from a violent interaction, and so on, and often a reduced fertility rate, leading ultimately to a reduction in densities. In nonhuman organisms, this generally results in a chaotic or sometimes regular oscillation around the environment's "carrying capacity"—that is, the total population size that the environment's available resources exploited by existing "technologies" can support.[8]

One of the most impressive things about human ingenuity is our ability to increase the local environment's carrying capacity, at least temporarily. We can produce food in vast quantities and ship it easily to places where little food grows. We can construct a mind-boggling ar-

*As it is, the discipline of ecology is certainly more complex—predicting how ecosystems composed of millions of species interacting with one another and their physical environment will change makes rocket science look like following a lasagna recipe.

ray of buildings with heating, cooling, and other amenities that shield us from weather extremes. We harness and store energy in a variety of forms—from body fat to liquid petroleum and electrical batteries—that assist in our transport needs. We can also devise clever technological ways to organize our complex societies to minimize aggressive interactions arising from dense living (think policing, video surveillance, financial transaction management, shopping centers). When accidents happen or when we get sick, we have medical care* that vastly increases our survival rate. These and other fruits of our ingenious labor allow most humans—especially the well-fed ones in countries like the United States and Australia—the luxury of not having to grow their own veggies, hunt their own meat, make their own tools, or even fight their own battles.

While our carrying capacities have grown largely as a function of our technological prowess, they cannot grow infinitely. Even a child can understand this principle. One can easily argue, too, that in both Australia and the United States, environmental carrying capacity has already been surpassed.[9] Despite over an order of magnitude difference in population size, Australia's appalling environmental record, marginal land productivity, and shortage of water, and the United States' staggering inequality in wealth distribution, which is on par with countries like Mozambique, Cameroon, and Uganda (see fig. 7.1), and a legacy of environmental damage and toxification, both countries are living well beyond their long-term means. Yet in both Australia and the United States, populations continue to rise, driven mainly by immigration.

An example demonstrates how the product of population size, consumption, and technological facilitation is the ultimate measure of carrying capacity. The amount of greenhouse gases that flow into the environment from energy use is a product of how many people there are, multiplied by the average energy use per person, in turn times a "technology" factor that measures the greenhouse gas yield of the energy-mobilizing system used (solar vs. coal or oil, carbon captured or not, Hummer vs. Prius, commuting by car vs. mass transit, insulation vs. air-conditioning, and so on). Society needs to know that it is all tied together: the more people there are, the more food society needs, and the

*In the United States, anyway, at least for the rich.

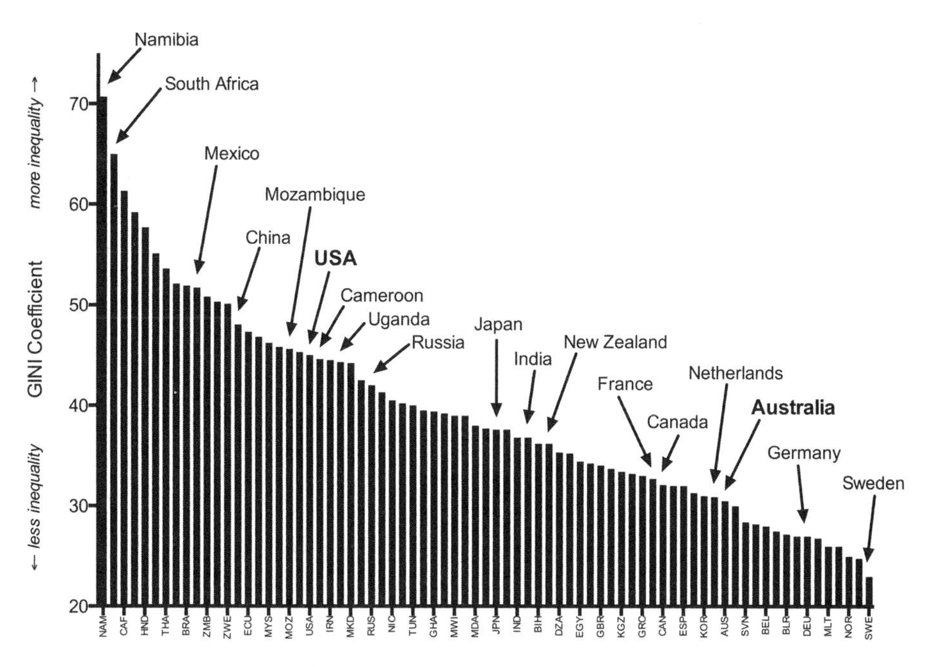

Economic inequality as measured by the Gini coefficient (named after the Italian statistician and sociologist, Corrado Gini) for a selection of countries. The coefficient measures income inequality, with a score of zero indicating equality in income across all members of society, and a score of 1 would mean that only one person possesses all a country's wealth. While not necessarily a perfect measure of wealth distribution, it is a reasonable index of the number of "haves" vs. "have-nots." Source: Central Intelligence Agency, *World Factbook*, https://www.cia.gov/library/publications/the-world-factbook/index.html.

agricultural system is a major emitter of greenhouse gases from fossil-fuel use, land-use practices, livestock production, and other factors. Thus, agriculture is a contributor to climate disruption, which in turn is a serious threat to food production.[10] With temperature and precipitation patterns now committed to more than a millennium of change,[11] including increased severity and frequency of storms, droughts, and floods (all already appearing), maintaining—let alone expanding—food production will be increasingly difficult. Agriculture itself is a leading cause of biodiversity loss[12] and thus the loss of the critical ecosystem services that biodiverstiy supplies to agriculture and other human enterprises.[13] Indeed, the human-caused hemorrhage of life-forms now under way[14]—the sixth great extinction event in Earth's

4.6 billion-year-long history—is certain to be accelerated by climate disruption but rivals it as a deadly danger to civilization.

Misunderstanding how demographic and environmental connections interact is unfortunately far too common, even among people who are interested in the problems associated with overpopulation. For instance, environmental reporter Fred Pearce[15] is convinced that overconsumption is a much larger contributor to environmental deterioration than overpopulation. This is roughly like being convinced that the length of a rectangle is a much larger contributor to its area than its width. History has shown that rapid population growth in most circumstances largely prevents the successful "development" of societies and retards increasing per capita consumption. What typically happens is that a nation's population grows rapidly for a while, followed by a period of slackening population growth and a concomitant increase in per capita consumption. That rapid growth of population and of consumption do not occur simultaneously is a small consolation, however, since the end result is a gigantic amount of consumption and the destruction of human life-support systems. Further population growth in super-consuming nations is super-dangerous.

China is the most obvious recent example of this as its *previously* skyrocketing population growth combined with its *current* skyrocketing growth in per capita consumption make it a champion in wrecking the environment on local, regional, and global levels.[16] Yet China's population growth is slated to end and even reverse by mid-century. India, on the other hand, is projected to add almost 350 million people by then and also seems bent on following the super-consumption highway. Similarly, an additional 80 million Americans by 2050 will enormously add to the already monumental US assault on the planetary systems that support humanity.

By contrast, consider the situation in sub-Saharan Africa, where more than 1.1 *billion* people are expected to be added to the present 920 million by 2050, more than doubling the population.[17] As Africans struggle to increase their inadequate levels of consumption, they will greatly increase the damage to the biowealth on which their societies utterly depend. We believe this puts an ethical demand on us rich folks to ease off on our consumption to give them some wiggle room.

This entire situation is made worse by nonlinearities in the population-consumption growth relationship.[18] Being clever and fundamentally lazy and greedy, human beings use the easiest and most accessible resources first. This means that the richest farmland was plowed first and the richest ores mined first. Now each additional person on average must be fed from more marginal land and less easily harvested wild fishes, use metals extracted from poorer, less accessible ores, and obtain energy from more difficult-to-access sources (e.g., deep-ocean drilling, fracking, and tar sands). Thus, each additional person disproportionately increases the potential and real destruction of environmental systems—the rate of destruction does not vary in direct proportion to the increase in population. The nonlinearities involved in resource extraction were dramatically underlined by the 2010 BP Deepwater Horizon blowout in the Gulf of Mexico. The first commercial oil well in the United States was drilled in Pennsylvania in 1859; it started at the ground surface and struck oil at 21 meters (69 feet). One hundred fifty years later, the Deepwater Horizon drill rig started a well for BP in the Macondo concession in the Gulf of Mexico; drilling began under almost 1.5 kilometers (1 mile) of water and had penetrated almost 5 kilometers (3 miles) below the seafloor when the explosion occurred, with its grim human[19] and environmental[20] consequences. Incidentally, BP is starting to explore for deep-water oil sources in the Great Australian Bight.[21] Australians should be very concerned.

As the population grows, keeping people supplied with consumer goods also releases more toxic compounds into the global environment. As we discussed previously, the toxification of Earth might be an even more dangerous trend than climate disruption or the extinction crisis, but it is increasingly clear that even the scientific community has not begun to address it properly.[22] People also should understand that population size is a major factor in the deterioration of the human epidemiological environment. The larger the human population, and the more hungry and thus immune-compromised people there are, the greater the chance of vast epidemics.[23] Almost everyone can relate to this on a personal level—have you ever been "run-down" and caught a cold? Think of several billion people "run-down" by hunger, and you get the picture.

Of course, many undereducated people think that the population can be kept growing by improving the "technology" factor, which includes sociopolitical issues of how consumption is supported and allocated. There is, of course, much room for improvement in both efficiency and equity. For instance, largely abandoning personal vehicles for commuting, and manipulating the economic system to reduce inequities, especially in food distribution, could greatly improve the human prospect. Of course, we could all choose to live like human beings did back in the Stone Age, but clearly that is not going to happen since there is no way even half of today's population could survive on Earth by subsistence farming and hunting/gathering. Put average American or Australian city-dwellers in the bush without supplies, and they would starve to death or die of exposure inside of two weeks. The history of claims that technological innovation will save us is instructive. When *The Population Bomb*[24] was published in 1968, the global population was 3.5 billion people, and P.R.E. was assured that technological innovation would allow society to give rich, fulfilling lives to 5 billion or more people. They would be fed by algae grown on sewage, whales herded in atolls, leaf protein, or the production from nuclear agro-industrial complexes.[25]

That, of course, never happened. The population now exceeds 7 billion, and the number of hungry and malnourished people today is roughly equivalent to Earth's entire human population when P.R.E. was born in 1932. As P.R.E. did after the *Population Bomb*, we challenge the population-growth enthusiasts to arrange to care (in a respectful, comfortable, and humane manner) for all extant human beings before providing more estimates of how easy it will be to feed, house, educate, and provide health services to billions more. It is sad how demographic-growth enthusiasts view the catastrophic expansion of human numbers. For instance, with the culture gap wide open, they celebrated the US population rocketing through 300 million people—well over double the number that could provide a safe and secure nation. In Australia, even when Kevin Rudd wrested the prime ministership from right-wing John Howard in 2007, he championed a "Big Australia" (36 to 40 million by 2050)—a target most Australians found distasteful.[26] Others bragged about the global human population size passing 7 billion, even though careful analysis[27] estimated this to be about 50% more people than could be supported permanently, even with today's level of misery for billions. Seven billion people would require several more Earths if everyone were to live like citizens of Australia or the United States.[28]

David Brooks,[29] generally viewed as one of the more thoughtful conservative pundits and holder of a degree in history from the University of Chicago, could be a poster boy for the culture gap. In 2012 he published a column on "the fertility implosion,"[30] joining a number of clueless European politicians, demographers, and pundits worried about a trend that could lead to a salutary population direction—shrinkage. They fear the aging of the population that inevitably occurs when population growth ends. All of Brooks's arguments have long been exposed as spherically senseless—uninformed from every viewpoint.[31] All one really needs to appreciate the silliness of fearing an aging population is realizing that the only humane way to avoid it is to keep the population growing *forever*. There are other reasons, too, why the "aging population" worry is also largely irrelevant—such misplaced concern completely ignores the other "dependent" component of the population: children. As the average age of a population increases when fertility declines, so does the number of children decline. The proportion of these dependents—retirees and children—relative to the "productive" (working) component of the population is known as the *dependency ratio*, which is a metric of the population generally considered to be dependent on the productivity of employed society.[32] Recent projections indicate that even under the business-as-usual scenario of declining fertility, the global dependency ratio remains remarkably constant at about 0.55 to 0.65 over the next century.[33] Therefore, the sociopolitical argument for encouraging high fertility rates to offset aging populations[34] that would otherwise put a strain on the productive (working) component of the population is mathematically, economically, and ecologically foolish.

Not only conservatives are relaxed about continuing population growth; many liberals also suffer from the same culture gap separating them from the realities of the world. Betsy Hartmann[35]—a professor at Hampshire College and director of its Population and Development Program—has many valid concerns about racism and the treatment of women in connection with population issues.[36] But her writings also clearly show that nothing in her education has allowed her to bridge the culture gap. She has degrees in Asian and development studies, disciplines traditionally isolated from the basics of the constraints of nature. Even the Australian "green" party (the Greens) is strongly focused on population growth, mainly via immigration.[37] Brooks, Hartmann, the Greens, and many other liberal commentators share their ignorance of how the world works with the majority of "educated" people, a problem partly traceable to the failure of schooling to put a proper emphasis on environmental science.[38] However, one might expect people to learn just a little before publishing misinformation.

Both culturally and genetically, human beings have always been small-group animals, evolved to deal with at most a few hundred individuals.[39] Even today that tends to be the number with whom most people associate, even in our gigantic, socially networked societies. Humanity is now suddenly, in ecological time, faced with an emergency requiring that it quickly redesign governance and economic systems that are both equitable and suitable for a global population of billions *and* sustainable on a finite planet. This likely cannot be done. As we pointed out, Earth is now so overpopulated that it would require many more planets to support permanently today's global population at the average American or Australian lifestyle, and even though several billion are now living in misery,[40] several billion more people are scheduled to be added to the population by mid-century.[41]

Humanity is exhausting its natural capital, including agricultural soils, fossil groundwater, the pollution-absorbing capacity of the atmosphere, and the biowealth that runs its life-support systems. It is disrupting the climate, spreading toxic chemicals from pole to pole, increasing the chances of vast epidemics, and risking nuclear war over resources, especially water and (in a monumental manifestation of stupidity) oil and, increasingly, access to agriculturally productive land. The scientific community predicts that at most a decade or two remains to revolutionize our energy-mobilizing systems, which are still mainly (more than 80%) dependent on fossil fuels, and revise our agriculture and water-handling systems to be flexible in the face of centuries of changing precipitation patterns. Any chance of growing enough food to give a decent diet to all of *today's* population requires success in these endeavors. Creating a just society, in which care for one another and our life-support systems move to the top of the political agenda, depends on massive social actions such as the Occupy movement—although even that worldwide phenomenon has now largely faded away. So perhaps the most basic question is how to transform our systems to help cushion the coming crash. How can we save the most lives and spare the most misery while transitioning to a scale of human enterprise that, with constant technological advance, will be able to persist long term into the future?[42]

8 Ignorance and Greed

Corporations are people, my friend . . . of course they are. Everything corporations earn ultimately goes to the people. Where do you think it goes? Whose pockets? Whose pockets? People's pockets. Human beings, my friend.

MITT ROMNEY (2011)

If you're jealous of those with more money, don't just sit there and complain. Do something to make more money yourself—spend less time drinking or smoking and socialising, and more time working.

GINA RINEHART, Australian mining magnate (heiress) and richest woman in the world, explaining that billionaires like herself are doing more than anyone to help the poor by investing their money (2012)[1]

In previous chapters, we outlined the history and current state of play of the environment in Australia and the United States, comparing the incentives, trajectories, and repercussions of dwindling biowealth in both countries. In this regard, these two great nations have a lot in common. Sensible people can see that that our consumptive lifestyles are a serious threat to our children's future, yet many people with highly selfish and short-term vested interests are diluting this message or are outright opponents of it. In many ways, this is why we chose to write this book—to inform people of the problems, identify those causing the most resistance to remediation, and spell out at least some ways we can overcome these impediments. Now it is time to be frank about the disciples of greed and ignorance.

Rotten reporting on environmental issues is not just a function of incompetence in the media. For example, casting doubt on the seriousness of climate disruption is now a major front in the US Republican[2] and, to some extent, the Australian[3] media's war on science.[4] It is grounded in an ideology that opposes regulation of industries that might limit the growth of profits, even if the regulations are required in order to avert future disasters. Those who try to mislead the public about the science of climate disruption are financed to large degree by the fossil-fuel industry and supported by propaganda from a fleet of conservative "think" (but in reality "fink"*) tanks and intellectual prostitutes. By "prostitute," we mean "someone hired for his or her [in this case, 'intellectual'] reputation to lend apparent credence to a worldview not supported by evidence." The anti-regulation ideology has been promulgated by a shameless group of pundits, some of whose careers trace back to being flacks for the tobacco industry, which tried (unsuccessfully) to persuade the public that evidence of smoking being harmful was "equivocal."[5]

Good—or bad, depending on your point of view—examples of corporate-backed misinformation abound. The outpourings of the Heartland Institute[6] are probably some of the best-known examples of how conservative vitriol in the United States is poisoning the minds of its citizens, although even their over-the-top virulence eventually proved self-harming[7] when they lost financial support by insinuating that climate scientists were akin to serial killers. A lesser-known and perhaps more insidious example involves an Australian industrial lobbyist, Alan Oxley. Former trade representative and ambassador for Australia, Oxley owns International Trade Strategies Global (ITS)[8] and is intimately associated with the World Growth Institute (WGI).[9] According to its website, ITS has "close associations" with several politically conservative US fink tanks, including the American Enterprise Institute, the Competitive Enterprise Institute, and the Heritage Foundation. ITS and WGI, along with its US allies-in-greed, are frequently involved in promoting industrial logging and oil palm and wood pulp plantations internationally—and have at times trod a thin line between reality and a major distortion of facts. Of course, funding anti-regulation propaganda organizations is dangerous enough, but buying elections is even worse. Between January 1, 2013, and August 31, 2014, the notorious Koch gang of environment wreckers were able to fund almost 44,000 television advertisements in an attempt to

*Defined as unpleasant and contemptible.

gain control of the US Senate and, thus, a free hand at profiting from further degrading the environment and continuing to immiserate the already poor.

In 2010 we were so concerned by the insidious activities of Oxley and his organizations that we joined a group of environmental scientists (led by our good friend and world-renowned tropical scientist Bill Laurance of the Smithsonian Tropical Research Institute and James Cook University) and wrote an open letter[10] about the scientific credibility of their claims, and how they are promoting the devastation of biowealth around the world. WGI, ITS, and Oxley frequently invoke "poverty alleviation" as a key justification for their advocacy of oil palm expansion and forest exploitation in developing nations. While true that these sectors can offer local employment, there are many examples in which local or indigenous communities have suffered from large-scale forest loss and disruption, have had their traditional land rights compromised, or have gained minimal economic benefits from the exploitation of their land and timber resources.[11] As we will discuss, there are many ways to alleviate poverty while improving the environmental situation (e.g., improving the rights and opportunities of women, making taxes more progressive) if the world economic system were not being run by and for the 1%. Let's face it, the rich are riding on the backs of the poor here,[12] as they have done since the golden age of murder, slavery, and plunder. Globally, well-financed fink tanks will do nearly anything to justify their rape of planet Earth now in obedience to the howl of their greed-driven funders and shareholders.

Happily, there are now efforts to counter the near-complete failure of the media and the political system even to begin to deal with the perfect storm of environmental problems facing humanity. Perhaps the bravest of these is embodied in the movie *Growthbusters*,[13] which actually had the nerve to point out that the emperor is indeed stark, staring, bollock naked—that physical growth of the economy is the disease, not the cure. Another recent release along these lines is *Greedy Lying Bastards*—a film exposing the corporate backing of climate-change denialism. And finally Naomi Orestes's great book *Merchants of Doubt* has been turned into a movie exposing the climate-denial prostitutes and their pimps. But all of these efforts are still too fragmentary and small scale to change the tide of public indifference and ignorance; paradoxically, they need to grow exponentially if the human enterprise as a whole is to undergo the necessary careful and humane shrinkage it sorely requires. For the public to support the idea that rescaling is necessary if we are to persist for a century or more, thousands of additional steps must be taken to close key parts of the culture gap in every society: such as the production of more documentaries like *Growthbusters* and making organizations such as the Post Growth Institute[14] and the Millennium Alliance for Humanity and the Biosphere more mainstream.[15]

But what about the two countries on which we are focusing—Australia and the United States? School systems, universities, and the media in both nations are doing an abysmal job of informing people about how the world works. They are still stuck in the original mission of public education—turning out large numbers of literate and numerate people who could be trained to be gunners in Napoleonic armies and repairers of primitive steam engines. They were taught—besides reading, writing, and arithmetic—obedience, punctuality, discipline, and other characteristics that the elite wanted to see in an industrial workforce. Thinking for themselves or "out of the box" was always a "no-no" and still largely is outside of the natural sciences. Even many of the most critical topics for citizens to understand are rarely covered. For instance, the single-most important thing that human beings do is acquiring their food. Nonetheless, an example of educational failure can be seen by simply asking people where their food comes from—the standard answer is "the supermarket." In 2010 a scientific literacy survey of the Australian general public found that about 30% believed that humans and dinosaurs had coexisted on Earth; one-third also thought it took one day for the Earth to revolve around the sun.[16] Worse still, the majority of "educated" people in both nations would be hard-pressed to discuss the role of transportation in the food system, the scale of the fossil-fuel contribution to agriculture, the importance of plant genetics and the genetic diversity of crops to our food supply, the role of agriculture in climate disruption, and so on. Of course anything vaguely covering crucial issues such as how corporations wield power, how "free markets" have never been significantly involved in "development," and how political systems actually work is completely out-of-bounds. This can be seen by the amusing lies told in American high school "civics" courses that pretend to tell us how the US government works, but tend to ignore issues like the power of corporations and the buying of politicians by them and other elements of the "elite" who actually governs the nation.

One might hope that modern electronic media can inform the public through adult education. So far the results do not look promising. The coverage of even prominent news events tends to be appalling in the mainstream media, the worst almost beyond belief. For example, those who watch the Murdochian *Fox* (dubbed "Faux") "*News*" anti-environmental propaganda network actually ended up knowing less about the world than those who watched no television news at all.[17] In Australia, even the better (non-Murdoch) newspapers carry nonsensical op-eds assuring us, for example, that "nimble markets"[18] will solve the looming food problems, even though over the last four decades they have not even provided decent diets to several billion people. Even the competitor media outlets are in imminent danger of libertarianism—one of the richest woman on the planet, Australian Gina Rinehart (who owes her billions to the country's mineral resources), has taken over a sizable portion of Fairfax Media shares (which publishes, among others, the *Sydney Morning Herald* and the *Melbourne Age*) and perpetuates her own propaganda against liberal policies in Australia. Thankfully, there are a few lonely voices of reason in Australian journalism that are bucking the libertarian trend. The irreverent *Crikey*[19] is a piercing online news source, and university-run *The Conversation*[20] commissions all of its articles from specialist academics across Australia and the world. There is also the independent *New Matilda*,[21] and now the *Guardian* has recently started an Australian version of their left-leaning online and print newspaper. Sadly, even the best liberal journals in the United States, such as the *Nation* and the *New York Review of Books*, have generally been weak and slow on environmental issues (especially on the role of overpopulation), although this has been improving. It is difficult to imagine any major future problem in the international arena that will not have a strong environmental component; nevertheless, the record of *Foreign Affairs* (and most political science journals) in this area has yet to reach even the standard of pathetic.

Despite these promising trends, the purveyors of environmental disinformation in wealthy nations have been stunningly successful. The case of Bjørn Lomborg, a minor Danish statistician in the Copenhagen Business School, is instructive. Lomborg has consistently tried to downplay environmental problems. His 2001 book, *The Skeptical Environmentalist*,[22] gained much momentum because it was published by Cambridge University Press. Being published by a reputable academic publisher gave it the aura of peer-reviewed work, even though it had clearly not been vetted by any environmental scientists. It eventually did receive peer review—mainly in the form of consistently scathing analyses in scientific journals.[23] Chris Lehmann, deputy editor of the *Washington Post Book World*, said that his newspaper had been fooled by the book because it was published by Cambridge University Press. Widespread condemnation and direct protests of the book eventually led to an investigation by the Committee on Scientific Dishonesty of the Danish Research Agency. The committee's decision, announced in January 2003, was that Lomborg's book violated Danish standards of scientific practice and met the criteria for "scientific dishonesty." It later retracted the "dishonesty" charge because Lomborg is not a scientist. We would welcome similar investigations into other books by purported scientists who commit what appears to us to be scientific dishonesty—books like Ian Plimer's *Heaven and Earth*,[24] Garth Paltridge's *The Climate Caper*,* and S. Fred Singer and Dennis Avery's *Unstoppable Global Warming*, to name some of the more egregious tomes. Should the US National Academy of Sciences and the Australian Academy of Science take up the challenge? We believe they should, and universities employing dishonest scientists should have internal review structures capable of sacking the worst offenders of scientific fraud and public deception.

Back to that book rotting in Denmark. The *Economist* attacked the Danish Committee, with its deputy editor, Clive Crook, saying that the Danish decision "offers nobody any reason to change their minds on Lomborg's books." After the ruling and all the scientific reviews, the *Economist*'s editor still stated that Lomborg's book, replete with elementary statistical errors, was an "outstanding statistical study." Of course, the *Economist*'s incompetence was not restricted to environmental science. As Sir Partha Dasgupta, one of the world's most distinguished economists, wrote, even the magazine's economic analysis "is rarely above the sophomoric."

*With a foreword by the, shall we say, "colorful" Christopher Monckton, this book can be shelved in the "sham" section of the library without having to read a word.

The point here is not to criticize Lomborg's (and his ilk's) meretricious book further (that has been done thoroughly by many others[25]), but to get an inside glimpse of how powerful interests can command distortion of the news in widely read and politically influential publications. The goal of the *Economist* and similar publications is often not to record an approximation of the truth, but to push a plutocratic political agenda. Amusingly, in June 2003 the Cambridge University Press bragged in an advertisement in the *New York Review of Books* that Lomborg had won the Julian L. Simon Memorial Award from a notorious right-wing lobbying organization in Washington, DC, the Competitive Enterprise Institute. The prize was named after Lomborg's hero, the late Julian Simon, a professor of mail-order marketing. Simon's scientific acumen can be judged by his stated beliefs. In 1980 he claimed, "Even the total weight of the earth is not a theoretical limit to the amount of copper that might be available to earthlings in the future. . . . Only the total weight of the universe . . . would be such a theoretical limit . . . because copper can be made from other metals." In 1994 Simon announced: "We now have in our hands—in our libraries really—the technology to feed, clothe, and supply energy to an ever-growing population for the next 7 billion years." Needless to say, long before such time had passed, the human population, growing at even a thousandth the rate when Simon wrote, would far exceed the number of elementary particles in the universe. Lomborg* richly deserved the Simon honor. We are not so sure about his reported[26] annual salary of $775,000—but then certain professions have often paid well.

Despite the environmental pooh-poohing of the plutocrats and their champions, most critical issues in today's world have a fundamental, indeed often overwhelming, environmental element to them. Two of the most obviously "environmental" threats to civilization—climate disruption and toxification of the planet—are clearly global in scale and import. To avert disaster will obviously require coordinated action by at least the major players—most prominently by the two most overpopulated and overconsumptive nations, China and the United

*As our book went to press, Lomborg was allocated AU$4 million by the right-wing Australian federal government to set up an Australian node of his so-called "Copenhagen Consensus Center." Initially earmarked to be based at the University of Western Australia in Perth, that university ultimately caved in to staff and public pressure and reneged on hosting Lomborg's Center. Lomborg and his Australian government supporters are now searching for a new host university.

States. But if we include the environmental consequences of wars, the deterioration of the epidemiological environment, and what security expert Professor Michael Klare describes as a "race for what's left" of resources ranging from minerals to farmland,[27] one can see that departments of state or their equivalent need to catch up with intelligence agencies and military establishments. They, too, must become knowledgeable about the environmental dimensions of national security, hopefully to help translate them into issues of international security. No country can possibly be secure while these critical issues remain unaddressed. Even the mighty US military recognizes such threats as climate change to its national security.[28]

Some of the most pressing environmental issues are already on the international agenda. Chinese dissident Chen Guangcheng—a critic of abuses in the administration of China's one-child family policy and now a denizen of US right-wing fink tanks—caused a nearly month-long diplomatic fuss that tested US-China relations. The difference between the attitudes of the Chinese government, trying to deal with its problems of vast overpopulation, and those of the US government, trying to ignore them, was palpable. China has had problems with its one-child family policy, which, among other things, has led to forced late-term abortions. There are better ways to achieve the low Chinese birth rates essential for both China and for the world.[29] Unhappily though, complaints about the Chinese policy have been accompanied by gibbering nonsense about the need for population growth to provide China with prosperity and spherically senseless statements about the aging of the Chinese population as growth slows and stops.[30] As we discussed earlier, aging of the population is inevitable unless the population can grow forever, and it tends to be balanced out by a declining number of young dependents.[31] In addition, virtually every other problem associated with changing age composition can be easily dealt with by appropriate long-term planning. Meanwhile, not surprisingly, Chen has ended up at an anti-abortion, anti-gay, right-wing fink tank.[32]

Recent demographic modeling shows that even under a (obviously fantastical) worldwide one-child policy implemented gradually until it was global by 2100, there would be roughly the same number of people alive on the planet in 2100 as there are today.[33] Moreover, even large-scale catastrophes such as world wars or global pandemics* would

*Grim thought experiments only.

have little effect on the final population projections to the year 2100. A specific example illustrates the massive demographic momentum that the human population now has. Let's take the three worst human mortality events in recent history—the First World War (15 million dead), the Spanish flu (50 million dead*), and the Second World War (66 million dead)—and combine them (131 million dead in total).[34] This number represents about 5% of the total world population alive after the Second World War. If a proportionally similar event happened again in, say, 2050 (i.e., a "Third World War"), this would equate to about 500 million deaths based on median projections of the world population. Spread over five years, this massive, dreadful, and unfathomable loss of human life barely makes a blip on the projections to 2100.[35] The implications are sobering—even mass death, epidemic disease, and morally questionable fertility control are not going to give rise to a substantially smaller world population any time soon; it will likely take more than a century, not decades. That said, the long-term benefits of reducing population size globally are without question, and even meaningful population reduction (i.e., hundreds of millions) through humanely reducing fertility rates and prevention of millions of unwanted births are possible by the end of this century,[36] if only we can teach our leaders some sense regarding good population planning. For example, if the world could achieve an average fertility of two children per woman by 2020, there would be over 770 million fewer people to feed by 2050 compared to a scenario without global-scale family planning.[37] Is this achievement plausible? Yes. Is it likely? We doubt global society will suddenly come to its collective senses by 2020 to implement such an important change.

Such differences in dealing with one of the two main drivers of environmental destruction (China with strong measures, Australia with scientists debating growthmaniacs, the United States totally ignoring the big picture with right-wing nutters obsessed with ending legal access to contraception and abortion) seem sure to cause more diplomatic difficulties in the near future as large population sizes generate rising competitive demand for disappearing resources and the dwindling capacity of vital environmental sinks. The other main driver, per capita consumption (both its scale and what is consumed), is already causing horrendous international problems as people die in Afghani-

*Estimates vary, but this figure is probably at the higher end.

stan, Iraq, and elsewhere in resource wars and local violence over fossil fuels that should not be burned. Environmentally, the US military contributes gigantically to climate disruption by burning about as much oil as Nigeria's 175 million people and causing much more burning by others, because the American military is used first and foremost as a tool for keeping the United States supplied with imported oil.

In view of all this, why does the foreign-relations community largely neglect the critical environmental dimensions of international relations? One reason, of course, stems from a dismal education system that is largely to blame for the ubiquitous and profound ignorance of the environmental underpinnings of civilization. The symptoms are many, starting with the rarity of discussion of environmental issues in international relations journals. Recently though, one of the premier journals already mentioned, *Foreign Affairs*, apparently thought that it could cover the environment by publishing an unreviewed, incompetent article[38] pooh-poohing environmental threats by someone with no credentials in environmental science. Besides failing to mention many of the dimensions of the global environmental predicament, the article showed deep confusion about those it did discuss. It serves as another example of how easily "educated" people can be misled about an area of science that even today is largely neglected by school systems, including at the university level. The author was none other than Bjørn Lomborg.

In such a situation, several things are obvious. First, the environmental science content of education in international studies needs to be greatly expanded. This should not be restricted to programs where the core component is "environmental"; rather, environmental education should become the backbone of law, political science, and business, among other subjects. This may be a considerable challenge for many colleges and universities that are weak in environmental sciences, but it is a challenge that can be met. There is excellent literature out there ranging from texts to review articles. Any competent scholar given a semester to retool should be able to grasp the most basic issues[39] and be able to introduce them to classes within her discipline. Second, help should be delivered to journals that publish in those areas to assure that they can get competent peer reviewing for articles with environmental content. Third, university faculties should try to assure that university presses are required to set up peer-review systems for whatever goes out under the university's name. It is noteworthy that faculty members at Cambridge tried and failed in that regard over the Lomborg fiasco, to the detriment of qualified authors who publish with that publisher (and, admittedly, they do produce many quality publications by well-respected scientists). Most important, those professionally involved in international affairs should institute programs to upgrade their understanding of the human predicament.

Journals and magazines such as *Foreign Affairs* and the *Economist* should publish regular articles and book reviews on pertinent topics in the international dimensions of environmental science. Of course, our discussion of this topic assumes that these journals can be partially freed from the mauling influence of industry money and publish peer-reviewed information. They could avoid being so easily scammed by poseurs by consulting with members of professional science organizations such as the US National Academy of Sciences, the United Kingdom's Royal Society, or the Australian Academy of Science. This would not guarantee that all articles would be of high quality—for example, an ex-president of the National Academy of Sciences, physicist Frederick Seitz, became a notorious intellectual prostitute working for the tobacco and oil industries.[40] Of course, environmental scientists, too, can sometimes suffer from greed and have political agendas—but fortunately that agenda is often trying to save a global environment that will support everyone's descendants in health and comfort. Sadly, on occasion they may selectively cite data or otherwise attempt to strengthen their cases illegitimately. But there is a difference between most of them and anti-environmental "brownlash" hacks like Seitz. Of course, conspiracy theorists could accuse us of being secret consultants for the condom industry, but they might have a difficult time finding the evidence to support their suspicions.

The scientific community is large and diverse, and the penalties for scientists doing bad science are severe: Seitz lost his reputation as a scientist when he was involved in publishing a paper forged to look like it was from the *Proceedings of the National Academy of Sciences of the USA*.* Scientists must subject their work to peer review if they are to maintain their reputations in the community, and their reputations are vital to getting the rewards that the community has to offer (such

*A forgery so egregious that the National Academy of Sciences issued a news release saying it was a deliberate attempt to mislead: www.desmogblog.com/oregon-petition.

as keeping one's job!). The fallacy among some non-scientists is that we collude on issues to dupe the wider public—as we discussed previously, nothing could be further from the truth. Believe us, if we attempt to publish anything without serious empirical evidence, we are instantly leapt upon by thousands of our colleagues baying for figurative blood. Scientists are not, contrary to popular opinion, friendly to one another when it comes to communicating our work. Even then, scientists do not always have to be right, but they have got to be honest, and the system mostly keeps them that way. But peer review and risk to reputation obviously do not concern some organizations or even some prominent individuals claiming to be scientists.

In light of this, it is sad that there is so little sensible discussion of, and even less action on, ethical issues in our societies. Just consider some of those that surround environmental problems.[41] One of the most pressing is that of intergenerational equity. What sort of world should we leave to our children and generations beyond them? Is it ethical, for example, to exploit resources unsustainably now on the assumption that our descendants will be so rich and technologically skilled that they will easily find substitutes? What about the *intra*generational issues? Different societies have made different contributions in the past to the atmospheric load of greenhouse gases. How past contributions should be weighed against current emissions has long been a bone of contention in negotiations between rich and poor countries, and is still without resolution. Indeed, the ethics of climate disruption are generally vexed because it is clear that, at least in the early stages of change, the poorest people are likely to bear the brunt of the problem. Is it ethical for any society to not try to provide universal health care (as does Australia* and only recently has the United States moved slightly in that direction), but rather to use a smokescreen of twaddle about markets and freedom simply to enrich the plutocrats who own insurance companies (as to a large degree the United States still does)?

It is no surprise, therefore, that a major factor in the failure of the United States or Australia to take any leadership position in dealing with the human predicament is the growing influence of an almost to-

*Australia currently has the world's sixth longest life expectancy (female: 85 years; male: 80 years) compared to 190 other nations, due at least in part to its universal health care policy. The United States has the thirty-eighth longest life expectancy (female: 81 years; male: 76 years) (source: United Nations, *World Population Prospects: The 2006 Revision* (New York: United Nations, 2007).

tally parasitic sector of the economy—the financial "industry" (since this sector produces nothing, one questions how a collection of rent-seekers could ever be considered an "industry"). Most efforts of this collection of manipulators focus primarily on enriching themselves and transferring wealth from the poor to the rich—otherwise known as the "Hood Robin effect." But even when it is functioning in the way that many analysts claim it should, with the capital markets and banks efficiently providing access to funds for investment and contributing to capital accumulation, the sector is doing a vast disservice to society. Basically, they fuel the growth of the physical economy, especially of the already rich, when preserving civilization demands a *shrinkage* of that economy. Sadly, many people fail to realize that the "standard" neoclassical economists' goal of 3.5% perpetual annual growth implies a biophysical impossibility—an economy more than thirty-two times as large as today's in a century. Even a middle-school child can easily conclude that even in the short term, such growth is a recipe for catastrophe—considering nonlinearities, it means much more than doubling humanity's destructive impact on its life-support systems in just twenty years.

Economists themselves—and we are talking about the discipline here (not just some dissident leftists)—know that gross domestic product (GDP) and its close cousin, gross national income (GNI), are poor measures of "wealth." We can think of GDP more as a speedometer, for it really measures the speed with which an economy is contributing to the generation of goods and services. It does not measure, for instance, the built-up capital in our society's infrastructure or the depletion of biowealth such as forests and mined resources. In fact, a major oil spill generally increases GDP because of the employment created to clean it up, whereas growing vegetables in the garden to feed the family is not included at all because the goods are not "traded" in the standard market.[42] Nor does GDP account for wealth disparity, so even though most people might be poor, the existence of a few fat cats can actually inflate GDP. Clearly something is amiss here. David Korten, author of *Agenda for a New Economy*, recently wrote that "both [US political parties] focus on growing GDP, ignoring the reality that under the regime of Wall Street rule, the benefits of GDP growth over the past several decades have gone almost exclusively to the 1%—with dire consequences for democracy and the health of the social and natural capital on which true prosperity depends."[43] So out of touch with reality is the GDP metric that the World Bank has endeavored to come up with better ones that attempt to measure *wealth*. One possible advance, although still falling short of measuring true wealth, is "total wealth" (TW), which is the present (discounted) value of future consumption that is "sustainable."[44] In other words, the TW metric tries to take into account what wealth a country possesses now, minus the damage it does to its non-renewable stock of wealth currently under unsustainable exploitation. By including the "stock" of existing capital, which encompasses natural capital (a major component of which is biowealth), economic policies based on this metric would potentially be in a far better position to ensure the long-term sustainability of society. Another candidate to replace the GDP is the Genuine Progress Indicator (GPI), which uses the base data to calculate GDP but adjusts them with other aspects ignored by the GDP, such as income distribution, environmental costs, crime, pollution, and volunteer work.[45] As such, the GPI better approximates sustainable economic welfare, but, unfortunately, so far it has only been calculated for a handful of countries.[46] If we could only get rid of the GDP-growth mantra spewed daily by politicians and media personalities alike, and get them to focus on measuring true wealth in their daily reports or speeches, we might at least have a viable way forward for benchmarking a society's real prosperity.

The connection between the growth insanity and the consumerism promoted by the dominant social paradigm in the United States, and increasingly in Australia, can perhaps be seen in the push for uncontrolled consumption that is a feature of the period toward the end of the year, anticipating the American Thanksgiving celebration in late November and the Christmas holiday. The season has changed from one originally celebrating religious myths, to one celebrating spending and gifts of "stuff." In the United States leading up to Thanksgiving, the mainstream media focuses on the extended opening hours of stores and the influence of shoppers on the "health" of the economy, featuring the unwise commentary of one-dimensional, growth-addicted economists. Interviews with those camping out overnight to be early in line to get bargains on such things as large televisions often contrast with those of homeless, unemployed people being given a free meal. Never is a word heard about the structural features of society that cause the homelessness—and much worse for billions outside of the rich world.

Rarely a word is said about the present and ultimate environmental consequences of growthmania.

The US financial industry is part and parcel of the government, and in that position has been largely responsible for the much-discussed redistribution of wealth from the poor and middle classes to the richest 1% of the US or Australian population: the super-rich. That in turn has reduced the knowledge and ability of the remaining 99% to protect themselves, their nations, and the world from the growing environmental threats that are rarely recognized by the ecologically ignorant money-grubbers who run the financial sector. Some of the ill-gotten financial gains of these parasites are showered on their well-paid agents in the US federal government to ensure that their licenses to steal remain endorsed and valid. For instance, between 2002 and September 2008, the financial sector gave over $1.1 billion to congressional candidates—somewhat more going to Republicans than Democrats.[47] That amount omits many hundreds of millions given to charities designated by leading politicians, and spent on lobbying for them, and the huge salaries and soft positions arranged for the agents when they leave government "service." In the 2008 US election cycle, with the bailout bill coming up, the financial services sector (which, to be fair, includes insurance companies, some of which, unlike Wall Street, serve some social good) gave presidential candidates Barack Obama $22.5 million, Hillary Clinton $21.5 million, and John McCain $19.6 million.

Key members of the US Congress also were great beneficiaries of funds from the financial services sector. Lynn Turner, who had been chief accountant for the Securities and Exchange Commission, summed up the corrupt mess: "If you trace the movement of Wall Street money through Washington, it pretty well tells the story behind this and any other piece of legislation. The way Washington works, it is the lobbyists and the executives who hire them that get what they want. And it is the taxpayer who usually ends up getting fleeced."[48] This would be bad enough if the financial sector made substantial contributions toward sustainable shrinkage, but instead they are a major factor in wrecking the world our descendants will inherit. In addition, they managed to reap vast rewards for being directly responsible for destroying the lives of millions of citizens. Their $22.5 million to Obama assured that they would suffer no appropriate punishment for their criminal acts. The industry should be largely shut down, its remains overhauled so that everyone can get honest banking services and financial advice (there are many admirable individuals in the business who can supply those), and the rich parasites who run Wall Street turned loose to take up less-destructive occupations for which they are well suited, as perhaps gangsters or conmen,* or for those we would trust around children, something positive such as teaching middle-school mathematics.

In the United States, the plutocrats are becoming increasingly powerful, pushing the country further and further toward disaster finances, just as they are increasing the efficiency of redistributing wealth from poor to rich, which is pushing society toward a total plutocracy. A recent stunning step in that process was the 2010 *Citizens United* case,[49] decided by a Supreme Court larded by wealthy presidents with enough agents of the rich and unprincipled to form a majority. Basically, the decision declared that corporations were "individuals" just like real people, and therefore entitled to freedom of speech. This was followed by the decision of the plutocrat conservative majority's 5–4 *McCutcheon* decision in 2014, which as the minority dissent accurately states:

> Taken together with *Citizens United v. Federal Election Commission*, today's decision eviscerates our Nation's campaign finance laws, leaving a remnant incapable of dealing with the grave problems of democratic legitimacy that those laws were intended to resolve.[50]

The two decisions reinforce the system of legalized bribery by which the United States is governed and almost guarantees further concentration of wealth and power in the hands of the few, and a few who are generally neither capable nor inclined to provide leadership in solving the horrendous social and environmental elements of the human predicament. These decisions add to the obscene influence of the likes of the Koch brothers, make a strong argument for policies of capping and redistributing wealth, and (once again) for *getting the money out of politics*. Increasingly, the United States is being converted into a

*To meet some of the ignorant disciples from academic departments of economics who support the criminals, we highly recommend the documentary *Inside Job*, available at Netflix (www.netflix.com).

plutocracy, buoyed by the myths of outdated macroeconomics—into what economist James Galbraith has coined a "predator state."[51]

The efficacy of the sector's efforts combined with those of the fossil-fuel industry, the military-industrial complex, the Chamber of Commerce, and other predatory growthmanic elements in society were well illustrated by the subsequent near-total failure of the Obama administration to deal with the major environmental issues facing the nation. That of climate disruption was widely recognized, and the scientific community, environmental nongovernment organizations, knowledgeable people within the administration, and educated members of the public were unable, despite determined attempts, to keep the president from giving in to his ignorant economic and political advisers and supporting, among other things, an insane "drill, baby, drill" approach to US fossil-fuel policies.[52] Of course, much of this can be traced to the US Congress, where, for example, the notorious senator for the oil and gas industry, Oklahoma's Jim Inhofe, who declared climate disruption a hoax, as of 2013 has received over $1 million of fossil-fuel industry money since 1999. Small wonder he works busily for the petroleum industry when he is not trying to abridge the rights of American citizens.[53]

Most of this chapter has been devoted to the US situation, mainly because of its massive economy (still, but not for long, the biggest in the world), wealth disparity, political extremism and insanity, and hyper-religiosity (more on that later). This is not to say that Australia does not have its own problems on this front, with the political attitudes and trends in the United States often being sycophantically imitated by many Australian politicians, more so now than ever before in Australian history. While Australia's Gini coefficient of wealth disparity is closer to the likes of the Netherlands, Canada, and Germany, its political extremism is definitely increasing. Even the official recognition that climate disruption was an issue worth considering did not arise until 2007 with the defeat of the Coalition government by Labor, and the appointment of the new prime minister, Kevin Rudd. At the time of his election, Rudd immediately ratified the Kyoto Protocol, which the former Coalition prime minister, John Howard, flatly refused to do during his entire three terms in office. The new government also quickly set up a Department of Climate Change—an office that up until 2007 was considered laughable in Canberra.

Since Labor's election in 2007, however, the treaty-signing and new-department-creating Labor government had done a few things, although vastly inadequate, of worth in the climate-disruption arena, all of which were overturned following the reelection of the Coalition in September 2013. Spearheaded by the current prime minister, Tony "the Monk"* Abbott, the party is bent on destroying everything to do with positive government action in the environment. Newly dubbed the environmental "Abbottoir" by many because of the party's unprecedented and all-out war against environmental protection, in less than six months the new government had already been singled out by the international community as one of the most hostile governments to environmentalism (and many human rights) that the modern era has ever witnessed.[54] Among the thirty or so egregious anti-environmental policy shifts the Coalition swiftly put into effect before the ink on the ballot papers barely had time to dry, some of the following suffice to illustrate the degree to which the party's ideology is firmly against any environmental protection.

The Abbottoir-led Coalition has, since election, tried to remove World Heritage status from Tasmanian forests so that they could be logged. Thankfully, the UNESCO World Heritage Committee took all of ten minutes to condemn the move,[55] even though Tasmanian forests are not yet out of the woods. No nation on the planet has heretofore attempted to delist a World Heritage Area. The Coalition also pressed hard to allow the dumping of 3 million tonnes of dredging spoil onto the World Heritage Great Barrier Reef to develop a coal-shipping terminal.[56] Although the spoil might now be dumped on land, the consequences for the Barrier Reef are not yet known. Almost immediately, the new government rolled back protection in Australia's national parks and protected areas, allowing logging, grazing, fishing, and hunting,[57] and vowed to oppose the creation of any more national parks.[58] They also attempted to remove the tax-deductible status of environmental groups and nongovernment organizations that work to protect Australia's unique biowealth[59] and tried to ban any boycotts of companies involved in environmentally damaging practices.[60] How that would be enforced is anyone's guess.

*So nicknamed because of his fateful surname, his beliefs in supernatural monsters, and the fact that he originally trained to be a priest in a Catholic seminary.

In the 2014 federal budget, the Coalition systematically slashed environment-related funding across the country, including major cutbacks to the federal Environment Department,[61] the National Landcare Network,[62] and scientific organizations responsible for environmental research.[63] As if to foreshadow the inevitable backlash, they had previously attempted to make the environment minister immune against future legal challenges to his decisions on environmentally damaging development projects,[64] and they axed a major funding scheme to the only independent legal entity available to limit environmentally destructive human development—the Environmental Defenders Office.[65] What foresight!

Other egregious environmental policies ensued. The government delayed enforcing Australia's new Illegal Logging Prohibition Act, which criminalizes the importation into Australia of illegally logged timber and any product made from illegally logged timber.[66] Most importantly, they managed to kill the infant carbon-pricing scheme designed by the Labor government less than two years after implementation[67]—Australia had set a world trend in financial disincentives to reduce national carbon emissions, and then set another for destroying it utterly. Swiftly after taking office in late 2013, the Abbott government also killed off Australia's first federal Department of Climate Change, refused to appoint a minister for science, and neglected to send any official Australian delegate to international climate-change discussions.

In many real ways, Tony Abbot represents the Australian version of the far-right American plutocratic and theocratic political momentum, which has vomited up many memorable gems like Inhofe and congresswoman Michele Bachmann. For example, in Opposition, Abbott repeatedly and defiantly promised to repeal Julia Gillard's Labor-government carbon price implemented in July 2012 (originally set at AU$23 per tonne of emitted carbon dioxide equivalents), despite its policy-entrenched framework and the absolute necessity to tax the heaviest polluters—it was one of the few election promises he actually kept. He was also adamant that the public-benefit mining tax—a modest, albeit flawed, levy imposed on the over-gorged mining industry profiting on the public good (mineral resources on public land)—would be axed. He succeeded in killing that, too, in September 2014.

Regardless of the flaws in the nation's carbon price system set up by the previous Labor government, it was arguably a step in the right direction. However, Abbott's increasingly weird right-wing condemnation of anything resembling environmental protection is reminiscent of the misnamed American Tea Party* movement, which ironically supports the richest of the rich via their vehement opposition to taxes. Under Abbott's Catholic cloak of ignorance, he is also a misogynist, describing a woman's virginity as the "the greatest gift you can give someone, the ultimate gift of giving."[68] His misogyny was even called out by then–Prime Minister Gillard in a famous 2012 speech when she said:

> I say to the Leader of the Opposition I will not be lectured about sexism and misogyny by this man. I will not. And the Government will not be lectured about sexism and misogyny by this man. Not now, not ever.[69]

While now a little more clandestine about his climate-disruption denialism (probably because it is ill-advised even in Australia's contemporary political arena), a denialist he remains:

> This is a government which is proposing to put at risk our manufacturing industry, to penalise struggling families, to make a tough situation worse for millions of households right around Australia. And for what? To make not a scrap of difference to the environment any time in the next 1000 years. (March 29, 2011; talking about the carbon price, which has since demonstrated almost no effect on personal finances.)[70]

> Whether carbon dioxide is quite the environmental villain that some people make it out to be is not yet proven. (March 15, 2011; again, in response to the carbon price.)[71]

> I don't think we can say that the science is settled here. (March 14, 2011; talking about climate change generally.)[72]

*The original Boston Tea Party was not a protest against taxes per se, but in opposition to taxes imposed by England's Parliament, where Americans' representation was only "indirect," as opposed to taxation through colonial legislative bodies.

> I think that in response to the IPCC alarmist—ah, in inverted commas—view, there've been quite a lot of other reputable scientific voices. Now not everyone agrees with Ian Plimer's position, but he is a highly credible scientist* and he has written what seems like a very well-argued book refuting most of the claims of the climate catastrophists. (December 2, 2009; talking about scientific charlatan Ian Plimer, whose book *Heaven and Earth* has been repeatedly and entirely debunked[73] by the scientific community.)[74]

There are many, many more similarly ludicrous Abbott comments, which others have shown abundant scientific evidence to the contrary.[75] Whatever the politics of the day requires him to say now, the Monk is obviously reticent to embrace any environmental policies that would be good for Australia. Here we have another plutocratic politician squarely within the back pocket of big business.

The Australian equivalent of Wall Street is not restricted to its financial industry per se; rather, the plutocratic tendencies of certain elements of government are bolstered predominantly by the fat cats of the country's major resource industries (mainly mining),[76] the big banks, and other major corporations. Although Australian political success is not as intimately tied to its donation portfolio as it is in the United States, there are worrying signs that this is heading in that direction. Traditionally, there has not been a culture of influential donations or the magnitude of corporate wealth in Australia to match the sort of financial lobbying seen in the United States. For the moment, the Australian Electoral Commission still requires political parties to disclose publically the total value of their received donations,[77] the total number of donors, individual donations more than AU$10,900, and the identity of all donors.† Unlike in the United States, where the average voter turnout percentage over the last twenty years ranges

*He is, in fact, a gold-mining geologist with no training in climatology. Even his peer-reviewed publication output in mining geology is minimal by Australian academic standards.

†Several attempts to harden these laws have failed but are ongoing. Proposed amendments include a prohibition of foreign donations, no anonymous donations above AU$50, and the creation of an exhaustive list of expenditures used to make election-funding claims.

between 35% and 55%, it is compulsory to vote in Australia. Regardless, one has to question how either Australia or the United States can call themselves "democracies" when big business can essentially buy their politicians of choice and corporations' various controls on the media and education are powerful. As the old saying goes, the United States and, to some degree, Australia have the best politicians money can buy. In our view, political donations of any kind should be made entirely illegal, and elections financed out of taxes.

A recent example demonstrates the slide toward big-business government is now gaining momentum in Australia. Billionaire Clive Palmer—a mineral-assets tycoon—flew onto the Australian political scene in 2013. Investing much of his own personal wealth to create his own party[78]—the cheekily entitled the Palmer United Party (PUP)—Palmer somehow managed to garner sizable portions of state and federal parliamentary and senate seats since the party's inception. No one has much of an idea what the PUPs aim to achieve in Australian politics, but between strange and ephemeral relationships with climate activists such as former US presidential candidate Al Gore,[79] falling asleep in the House of Representatives during question time,[80] and supporting the Coalition's dogged policy to ax the carbon and mining taxes,* Palmer is a loose cannon. That a billionaire can fund a party and gain real political power in Australia in less than two years is disconcerting, whether or not it challenges the establishment.

But the baneful influences of religions, corrupt politicians, out-of-control financial industries, and out-of-date political systems play out in both Australia and the United States against the more pervasive and less recognized cult—that of super-consumption, market norms, and growthmania. The first was recently emphasized for P.R.E. in a conversation with a bright, charming eleven-year-old son of a friend. When asked what he wanted when he grew up, he answered, "Many billions of dollars so I can have twenty homes." This led to a long discussion of the problems of managing so many domiciles, in which the boy simply assumed that money could solve them all by providing a pyramidal

*Palmer negotiated with the Coalition in mid-2014 to support the latter's bill to ax the mining tax. Such a reprehensible conflict of interest barely lasted the duration of the twenty-four-hour news cycle.

army of managers and watchers. His further discussion of goals and happiness focused virtually only on the means to buy the newest electronic gadgets and video games. P.R.E. was assured by the friend that this attitude was representative of the boy's peer group, something easily confirmed by general observations of consumer behavior, or by viewing the daily finale of the *Ellen DeGeneres Show*[81]—in which people given gifts worth a few thousand pretax dollars at least simulate going berserk with joy.

While insane population and consumption growthmania is generally expected among run-of-the-mill economists (the good ones* know better), they at least have an excuse of near-nonexistent training in how the biophysical world really works.[82] But how does one explain why the staff of *Science* magazine, one of the top scientific journals in the world, wrote the following misleading nonsense to introduce an issue on population?

> Today these demographic patterns spark concerns, not of a single explosion, but of "cluster bombs" in rapidly growing countries such as Nigeria and Pakistan, which are hobbled by poor governance and limited schooling capacity and already have huge numbers of poorly educated young adults without job prospects.[83]

The statement repeats the ancient error of focusing population problems on the poor, rather than the rich, and ignores, among other things, the global problems that are the most difficult to deal with consequences of continuing population growth. At any average level of affluence and available technologies, those difficulties are proportional to how many people there are altogether. The problems include, above all, global climate disruption, as more people add more greenhouse gases to the atmosphere, but also the looting of oceanic life for protein, emission of more toxic chemicals into the biosphere, more tensions and conflicts over access to oil and other resources, increased probability of lethal epidemics, and so on. With the partial exception of the

*Such as Ken Arrow, James Blignaut, Herman Daly, Partha Dasgupta, Larry Goulder, Quentin Grafton, Tom Kompas, Karl Goran Mäler, Charles Perrings, Steve Polasky, Joe Stiglitz, and many (but not enough) others.

epidemics,* these global problems are primarily generated by overpopulated rich nations and rich segments of poor nations, *not* Nigeria and Pakistan. Rich nations like Australia and the United States love to play the developing nation name-and-shame game.

The capitalist economic system traces back to the agricultural revolution, when families first became sedentary and were able to produce more food than they consumed. That laid the groundwork for a division of labor, great differentials of wealth and power (it's hard to accumulate stuff if you are a nomadic gatherer-hunter), and a period of unprecedented growth of the human enterprise. A combination of markets, private ownership, and the organization of corporations has led to enormous riches for the few and prosperity for a substantial minority of humanity. But a dark side of capitalism, collusion among capitalists (anticipated by Adam Smith†), and an insane belief that physical economies can grow forever—a myth that is endemic to politicians and second-rate economists in both Australia and the United States—have resulted in an extraordinarily threatening global environmental situation. The most immediate threat is to poor people and poor nations, although in the end the rich will also falter and crash. We will not go over the obvious critique of the impact of crazed capitalism on the environment here. Those interested in an academic discussion, there is an excellent one in William I. Robinson's *Global Capitalism and the Crisis of Humanity.*[84] Naomi Klein's recent *This Changes Everything: Capitalism vs. the Climate*[85] is a good read and makes many valuable observations, but it suffers from her lack of clear understanding of environmental science. What is certain is that the capitalist economic system as we know it is doomed; the main challenge for economists today is to work out how a non-growing system might be designed that could provide reasonably decent lives to, say, a billion or so people—and how to get there humanely. For a more sophisticated critique of the capital-

*Rapid transit systems are a major part of the deteriorating epidemiological environment and are primarily a rich-world phenomenon. But most of the other elements now contributing to the threat of vast epidemics can be traced more to conditions in poor countries.

†Eighteenth-century Scottish economist famous for his description of the "invisible hand" of the market that supposedly evolves as an emergent property of capitalism to the benefit of all society.

ist system and its promotion of "free markets," you might want to listen to that of Noam Chomsky.[86]

The bottom line is that both the United States and Australia are in the middle of a global emergency that is unrecognized by their leadership and by the vast majority of their populations. In this respect, they are in the same boat as the rest of the world, but with less excuse than most for their appalling inaction. For they have financed and created communities of environmental scientists far superior to those of most other nations, but then they have determinedly neglected their findings and advice, and have even attempted to gag these scientists.[87] It would be a sad indictment of our countries if our politicians, media, and corporations were merely ignorant of Earth's predicament; instead, it is entirely criminal that many *do* know and fail to respond. Is there any way this could be fixed? In the United States, and to an increasing degree in Australia, getting the money out of politics would be an excellent start.

9 Theocracy

At the bidding of a Peter the Hermit millions of men hurled themselves against the East; the words of an hallucinating enthusiast such as Mahomet created a force capable of triumphing over the Graeco-Roman world; an obscure monk like Luther bathed Europe in blood. The voice of Galileo or a Newton will never have the least echo among the masses. The inventions of genius hasten the march of civilization. The fanatics and the hallucinators create history.

GUSTAVE LE BON, *La psychologie des foules* (1895)

If the bankers, corporations, and greedy pundicrats were not enough, another environmentally painful feature of the culture of the United States—and to a growing extent Australia—is its excessive religiosity, especially the strength of right-wing Christian fundamentalism. Many of these people actually believe that the world was created some 6,000 years ago by a supernatural entity that is always metaphorically peeping through our bedroom (or, if they are evangelists, motel) windows. They know little or nothing about how we and our life-support systems coevolved, and they depend heavily on their god to solve our and their problems. The prominence of religious extremists, reminiscent of the ultra-conservative Taliban of Afghanistan or those Islamic State lunatics fighting for a seventh-century caliphate, is one of the most ominous signs that the United States, and perhaps Australia, will be politically unable even to start dealing with critical issues in the human predicament. One only has to note the steady descent started by Ronald Reagan (who had difficulty telling the difference between movies and reality) of the Republican Party into the outermost fringes of civil madness. At its most extreme, two of its recent major candidates in 2012, Rick Perry and Michele Bachmann, were actually connected to a fringe fundamentalist movement called Dominionism. Its basic tenet is that Christians should rule the world.[1] No, that is not a joke, over-exaggeration, or metaphor. They really mean it.

Perry's approach to environmental problems is exemplified by how, as governor of Texas, he dealt with the issue of the state's severe drought. On April 21, 2011, he issued an official proclamation that "the three-day period from Friday, April 22, 2011, to Sunday, April 24, 2011, as Days of Prayer for Rain in the State of Texas."[2] Perry's decision to declare a three-day session of prayer for rain, rather than to listen to the appeals of the Texas Forest Service to upgrade the fire-protection system, is a prime example of irrational plutotheocratic governance. As a result, thousands of these "rational utility maximizers," irrationally believing in a supernatural world that influences their lives, did not realize that the fossil-fuel industry was paying hundreds of millions of dollars to fool them about the risks of climate disruption, and they subsequently lost their homes to wildfires as a result. It was the worst wildfire season in Texas's history, its results attributed in part to people having made irrational choices in their voting. Rest assured, there will be ever more catastrophic wildfires in both the United States and Australia that no amount of prayer can prevent.

Perry's statements and actions are not just a sign of his religious beliefs, or perhaps his thin grasp of reality; they also reflect either a deep ignorance about how the world works or an immoral lack of caring about the fate of his children and more distant descendants. In 2010 he stated:

> I do believe that the issue of global warming has been politicized. I think there are a substantial number of scientists who have manipulated data so that they will have dollars rolling into their projects. I think we're seeing it almost weekly or even daily, scientists who are coming forward and questioning the original idea that man-made global warming is what is causing the climate to change. Yes, our climates change. They've been changing ever since the Earth was formed.[3]

This amazing statement* not only ignores that the "manipulated data" have been shown by every study to be imaginary, that "the sci-

*Perhaps not so amazing for the intellectually challenged governor who carries a gun whenever he jogs for "protection against snakes": nymag.com/daily/intelligencer/2010/04/texas_governor_rick_perry_jogs.html.

entists who are coming forward" exist only in his mind,[4] and that climate science (like almost all science) depends in part on a flow of funds into the research. The statement is also symptomatic of a tired accusation thrown at scientists who challenge the plutocrat's creed—that we somehow benefit financially from these perceived conspiracies. Just so the reader is clear, neither of us knows a single scientist who has ever become rich from public science funding. It does not happen. It likely never will. But it is part of the canon of those who live in a faith-based rather than an evidence-based world.

Michele Bachmann is also of that ilk. She is a marginally educated over-reproducer (five children) and clearly would reinforce dangerous pro-natalist trends in the United States:

> My record of supporting traditional marriage, family life and children, including those yet unborn, is unambiguous.[5]

Her Dominionist ideas include praise of those who have too many children and support of the nonsensical idea that the human population is actually shrinking.[6] How tenuous a grasp of reality can one person have? But Bachmann is ethical and self-restrained compared to the famous Arkansas couple Michelle Duggar and her husband, Jim Bob, who announced in late 2011 that they were expecting their twentieth child—sent (according to Michelle) by her imaginary deity.[7] They constitute a living plea for education, especially sex education, and are a monument to selfishness.

Social scientists have several hypotheses about why the United States is so plagued by religion. One is simply that the United States does not have an established or specially recognized religion. In nations with established religions, the religion gets mixed up with politics and government, and people quickly grow cynical about that intersection. Considering the obviously preposterous claims made by most religions, it is easy to see why people become cynical and therefore gravitate to a more generally evidence-based approach to the world. Some evidence for that can be seen in attitudes about perhaps the most powerful driver of environmental destruction—excessive childbearing. The United States has a total fertility rate (essentially completed family size) of 1.9, when an environmentally justifiable and ethical number would be more like 1.5–1.7. It is the same for Australia. In contrast, officially Catholic Italy is at 1.4, Spain at 1.3, Poland at 1.2; and when a pope wounded in an assassination attempt pleaded with Italians to repeal a liberalized abortion law in May 1981, more than two-thirds of the voters properly ignored him. An apparent exception to this pattern is France, which has a "high" total fertility rate of 2.0. But that is an exception that proves the rule that Catholics control their reproduction in view of that country's massive government child payments. Although about 80% of the French are listed as Catholic, France is in fact a deeply secular nation in which Catholicism is liberal to begin with and has been fading since the guillotine replaced the *auto-da-fé* during the French Revolution. Most French nominal Catholics go to church rarely, if at all, and care little for a sexually disoriented old man's views on human reproduction.

The situation in Poland is different, where more than four-fifths of the population actually claim to believe in a peeping-Tom deity. Its government supports freedom of religion but also gives special recognition to Catholicism in its constitution. Tending to follow the dictates of the Catholic hierarchy, the government makes it difficult for people to practice contraception. Nonetheless, most Poles seem happy to contravene the Pope's unethical rules* and limit their reproduction anyway, suggesting that despite their claimed beliefs, they are cynical about the purported divine commands. In northern Europe, with a predominantly Protestant religious history, and in long-featuring, established, or favored religions elsewhere, most nations have total fertility rates below 2.0 (Ireland is at 2.0). Even in India, on track to become the most populous nation on Earth, the Pope's henchmen are struggling to raise birth rates, in part to compensate for explosive growth of the Muslim population.[8]

While in no way near the religious fanaticism of US politics, Australia has its own growing theocratic tendencies. There is, of course, Tony "the Monk" Abbott, the Catholic climate-disruption denier. Recently, a front-bench member of Abbott's Liberal Party,† Senator Cory Bernardi—well known for his outrageous and religion-fueled conservative nonsense—equated gay marriage with bestiality.[9] The differ-

*We hope some of our commentary in this section will soon be obsolete. Pope Francis has just thrown his full weight into the battle to deal with climate disruption.

†A curious irony in Australia, where "Liberal" (uppercase *L*) is a major political party with the fewest liberal (lowercase *l*) policies.

ence though is how this remark was handled—in the United States, offensive and ridiculous claims regarding sex and marriage are often worn as a badge of pride by right-wing politicians. In Australia, Cory Bernardi was sacked from the Coalition's front bench (but not from the party) the day after his public display of spectacularly uninformed bigotry. Even the Monk found his comments outrageous.

Australia's version of the plutotheocratic Tea Party is the Family First Party, which "believes government controls over the lives of its citizens should be wound back and individuals should be encouraged to take more responsibility for their lives,"[10] and that "marriage is exclusively between a man and a woman."[11] They want government out of everyone's lives—except when it comes to sex, where they want it between the sheets with you. Family First is also vehemently denialist[12] when it comes to climate disruption:

> Many hundreds of eminent scientists have strongly criticised both the "climate change doctrine" and the predictions made by the International Panel on Climate Change. Claims that "there is a scientific consensus" and "the science is settled" are not true.
>
> Carbon dioxide is not a pollutant, it is plant food. The more crops can get of it the better they grow.

One almost has to laugh at the party's brazen lack of evidence, scientifically incorrect statements, and utterly backward policies. While not yet any real threat on the national or state political scene, and tame compared to the lunatic fringe in the United States, their mere presence is a worry for all Australians concerned about their future. Australia also fails to make any clear separation between church and state, with religious infiltration massive even in government-funded schools (and without parental consent).[13] In fact, C.J.A.B. had to argue with Governing Council members of his daughter's public school that students should have to "opt in" to church-provided propaganda programs, rather than being forced to "opt out" by signing forms in advance. Happily, he won that local battle, but nationally the dark demons of religiosity are gaining stronger footholds in Australia. The 2014 national budget laid out by Abbott's government was a clear sign that religious zealots are back in power. After cutting hundreds of millions of dollars from science and environmental programs,[14] based on the contrived rationale of a "budget crisis,"[15] the government planned to spend some $250 million on chaplains from religious organizations to be established in public schools. Chaplaincy programs are an unfortunate legacy in many state-run schools across Australia, but Abbott's budget mandated that the program should become entirely the bastion of religious teaching. The plan offended many and led one man, Ron Williams,* to challenge the move in the High Court. Although the High Court did not oppose chaplains in state schools, they did rule that federal money could not be used to exclude secular counselors instead. It was a small win for reason in an age of increasing Endarkenment. Immediately following the decision, Queensland Liberal Member of Parliament Andrew Laming, who helped start the chaplaincy initiative, slammed the ruling as "frivolous" and remarked:

> I look directly in the eyes of the loose alliance of Greens, gays and atheists who have mounted this continuous campaign against chaplaincy: you are clearly out of touch.[16]

The irony is palpable.

On the bright side, Australia does have a relatively new political party whose main platform is the regulation of Australia's population and the avoidance of the mess overpopulation causes—the Stable Population Party of Australia.[17] While not yet a major political force, their support base is growing and could easily rival the Greens within the next five to ten years. There is nothing even remotely similar in the United States, and, of course, it is clear that we do not have much time to change our ways profoundly.

Belief in a supernatural source of rules clearly can have a sad impact on demographics, but not in the ways commonly thought. Many think that the population problem is a Catholic one, but as we have just discussed, Catholics tend to perform reproductively very much like non-Catholics in similar economic situations. The self-destructive idea of population-growth competition is hardly confined to Catholics. Some Orthodox Israeli Jews are determined to try to outbreed the Arabs—many Israelis fear the extremist doctrines and behaviors of the Orthodox Jews more than their official Palestinian "enemies."

*While a heroic gesture that we applaud, Mr. Williams is the father of six children.

In the United States, the main "Catholic problem" is that of a comfortable, self-satisfied hierarchy, too many members of which are focused on controlling sexual behavior of others while they themselves are engaged in the sexual abuse of children or protecting other abusers among their colleagues. Sadly, the abused tend to be youngsters made especially vulnerable by exposure to their church's antiquated sexual dogmas—like most other religious education, in our view it is a form of child abuse. Interestingly and happily, there is a movement of "rogue" nuns attempting to reorient the hierarchy and the Church as a whole toward doing more to help people[18]—a small but important first step toward creating a force that will help build a humane, evidence-based society that can seek sustainability. Some groups of nuns, at least, have an admirable record of helping the helpless under extreme conditions in Africa and elsewhere.

It is one thing for people to hold crackpot beliefs about a supernatural realm—everyone is entitled to his or her own personal fantasies—yet it is quite another to take political action to try to impose imaginary commands on others. The Church's views on population, contraception, and abortion—the promotion of dangerous nonsense—are thought by many to be due simply to the hierarchy's fear of admitting that Protestant faiths have been right all along. Before Pope John Paul took office, he justified pursuing unethical policies on birth control because a reversal would threaten the idiotic doctrine of infallibility.[19] But of course, the hierarchy has long tried to control other people's behavior by making up stories about what is right and wrong sexually, and by targeting the helpless for their own, often sadistic, sexual pleasures. Their views now mesh nicely with the Far Right's ideology of keeping women in a subservient place in society, arrogating to themselves the role of controlling women's lives and reproductive choices.[20] Suppressing women has long been a tradition in the "great" henotheisms and monotheisms; because the practice helps keep birth rates high, it amounts to one more way that ignorant old men can keep their power without considering the greater consequences for civilization.

It would be unfair to claim that all religion has a negative influence on the environmental front, except to the degree that it keeps people from thinking for themselves and devolves responsibility in most cases to an unseen imaginary dictator. Some fundamentalist sects are beginning to focus on planetary stewardship and protecting the "creation,"[21] and the declining mainline religions also include protecting the environment among their social responsibilities. Even the Catholic Church does some good despite its immoral hierarchy and the totally expected (and millennium-old) sexual abuses perpetrated by some of its clergy and the actual reprehensible behavior of some "saints."[22] Catholic charities are sometimes the only source of succor for the poor. In any case, Bertrand Russell's cosmic teapot* or not, humanity's problems cannot be solved without the collaboration of the "faith community"[23]—no matter how much their ideas are divorced from reality. Unfortunately they make up too large a slice of humanity to allow them to be ignored or circumvented entirely. It is clear that even if a reasonable, universal education system could be constructed, humanity is unlikely to be largely freed of the curse of faith in the foreseeable future. Recent trends toward religiosity after humanity's brief affair with reason suggests that, if anything, religious extremism will continue to grow with rising human population density.

People of the Perry-Bachman-Abbott affliction are becoming increasingly powerful, pushing our countries further and further toward greed-driven politics justified by religion. Indeed, most economists—and essentially the entire economic profession—assume that the economy operates on a pattern of human behavior that has been demonstrated to be nonexistent. They assume that people are rational, fully informed of pertinent information and that they make their decisions on the basis of attempting to maximize their happiness ("utility" in economic jargon). In fact, people base their decisions very much on emotions—emotions heavily manipulated by powerful economic and faith-based interests—and all too rarely try to increase their happiness directly or (perhaps more importantly) indirectly by maintaining or enhancing the good of the social system in which they are embedded. Consider, for example, the Texans who voted for Rick Perry for governor, who used his position to enrich himself.[24]

Perry is certainly not a lone plutotheocrat—but he serves society well as a horrible example. Unfortunately, Perry represents an entire wing, possibly the majority, of the Republican Party. The Perry phe-

*Russell created this analogy of proving that a teapot does not orbit the sun between Earth and Mars to argue that the burden of proof should remain with the person making scientifically unfalsifiable claims, rather than shifting it to others as the religious tend to do: http://rationalwiki.org/wiki/Russell%27s_Teapot.

nomenon is also growing within Australia's political establishment. Bragging to be under the control of a mythical being, that wing wishes to destroy any power the government might have to protect the poor and serve the common good. It uses whatever influence it can wield to enrich the super-wealthy further and convince everyone that they represent a vengeful (or witless) deity to which the world must pledge allegiance or perish. There are many others like Perry promoting the Endarkenment. For instance, Republican member of the Science, Space, and Technology Committee of the House of Representatives, Paul Broun, has dismissed evolution, the Big Bang theory, and embryology as "lies straight from the pit of hell." A medical doctor who claims that, "as a scientist,"* he has data showing that the Earth is no older than 9,000 years and was created in six days, instead of the 4.6 billion years indicated by all the prevailing evidence. He claims that theories of the origins of the universe and evolution are "lies to try and keep me and all the folk that were taught that from understanding that they need a savior." The comments indicate again the quality of some members of that committee. Republican representative Todd Akin of Missouri earlier attracted attention to his misogynistic imbecility by stating that it was "really rare" for rape to result in pregnancy. "If it is legitimate rape, the female body has ways to try and shut that whole thing down," he prattled, while defending his position that abortion should be banned in all circumstances.[25] We have doubts that most politicians actually believe such childish and repugnant[26] nonsense propagated by religion, but to get elected they think they must pretend to. Since many people clearly do believe, it is important that believers somehow be united with those who orient to the world in an evidence-based manner to work in the cause of sustainability.

There is some progress in this direction, as some evangelicals are starting to express concern about preserving "the creation."[27] We are pleased by that, even though we know how both the creation and religion evolved.[28] In fact, we think we do need a spiritual approach to saving the world, one that is, as P.R.E. has said before, "quasi-religious." Of course, we do not mean that we condone all the attention paid to how imaginary creatures who care deeply about how and with whom and

*He has an undergraduate degree in chemistry—not exactly what most scientists would label a "scientist."

in what positions we have sex, who control who lives and who dies, and who condemn people to an afterlife of either agony or boredom. Instead, we want a society that features such beliefs to be replaced by one that cares deeply about how fairly its members are treated in the here and now and how carefully its life-support systems are maintained. We want ethical questions to be central to everyone's lives, with all people understanding the constraints set by scientific knowledge (no, there is not a hell) and by the limits of science (no, evolution does not tell you if having three kids is unethical). We claim it is ethically required for all people to work hard to move in that direction.

As we warned at the beginning of this book, we promised to be frank. We are both convinced that the adaptive value of living in a faith-based world is long over. In fact, prosperous democratic societies tend to be worse off when they are heavily influenced by religious belief. A higher proportion of people who believe in and worship a god is correlated with higher homicide, juvenile mortality, sexual disease, teen pregnancy, and abortion rates.[29] Policies today, especially in the United States, are seriously affected by fundamentalist Christian sects as well as by those that subscribe to more standard mythologies. This is maladaptive in a democracy where it is considered impolite to criticize religion—indeed, even to discuss someone else's beliefs, regardless of how silly.[30] To have a major cause of the human predicament above criticism could prove lethal to society. Interestingly, a professor of religion, Charles Kimball, put his assessment of this norm bluntly:

> We must quickly unlearn that lesson. Our collective failure to challenge presuppositions, think anew, and openly debate central religious concerns affecting society is a recipe for disaster.[31]

This is an essential point since the political clout of the fundamentalists, partly because of geography and numbers of believers and partly because of ideology, is tightly connected with the American pro-resource war and plutocratic, anti-science political postures. These have led to an inability of the United States to take a leadership role in preserving our life-supporting environment and in altering how we treat other human beings in ways that might make a sustainable society possible. Australia has an opportunity here to avoid the fundamentalist mess in which US politics have become mired, but so far the signs are not good. Now plagued by political parties espousing faith-based nonsense as policy, Australia too risks entering the Endarkenment.

So criticize we have and will. The fact is that there is no evidence whatsoever, not the tiniest shred, that there are supernatural forces that tell us what to do or that will save us when we screw up. The disaster scenes in which the television news reporter is told by a survivor that "God was protecting me" are a hangover from the pre-Enlightenment era. The reporter never asks, "Do you suppose god hated those whom she let be killed?" If someone tells you a god created the universe, ask him what created that god? Humanity has little time to get its act together, and spending time and effort involved with imaginary beings thought capable of altering the course of civilization utterly wastes that time. We are *way* past the point where *Homo sapiens* needs to grow up and live in evidence-based cultures. It is probably our only chance—remote as it seems.

10 Circling the Drain

For 200 years we've been conquering Nature.
Now we're beating it to death.

TOM MCMILLAN, former Canadian Minister of the Environment,
quoted in *The Greenhouse Trap* (1990) by F. Lyman[1]

Make no mistake—the scientific community is clear that humanity is in the midst of an unprecedented, slow-motion global emergency,[2] yet that emergency is barely recognized by most people or the politicians who are supposed to be representing us. When your accountant tells you to reduce expenditures, you do it or risk bankruptcy; when your electrician tells you the wiring in your house is dodgy, you replace it or risk your family dying in an avoidable fire; when your doctor tells you your cholesterol is too high, you cut back fat intake (and/or take cholesterol-reducing drugs) or risk a heart attack. Yet few with any real political or financial power heed the warnings of environmental scientists. It is not just a few of us either—globally, ecologists, conservation biologists, and environmental scientists are united in telling the world (for decades now) that growth in population and consumption cannot go on forever. They have been united in telling us if we do not clean up our planet, our life-support systems, which are declining rapidly, could ultimately fail. As we have discussed in previous chapters, that decline is manifest.

There are now well over 7 billion people on Earth, and median projections suggest that the population will grow to 10 billion or more by the end of the century.[3] Some analyses[4] indicate that with present technologies, Earth could only sustainably support indefinitely some 5 billion people under best-case scenarios, but assuming similar proportions of poverty and suffering as we have today.[5] Others imply that 5 billion could be too many.[6] As a result, humanity is entering that near-perfect storm of problems driven by overpopulation, overconsumption by the rich, gross inequalities, and the use of needlessly environmentally damaging technologies. The problems include the intertwined dilemmas of loss of the biowealth that runs human life-support systems; climate disruption; energy shortages; global toxification; alteration of critical biogeochemical cycles; shortages of water, soil, mineral resources, and farmland; increasing probability of vast epidemics; and the persistent specter of a civilization-destroying nuclear war[7] that could very well start over shortages of natural resources.

On the climate front alone, greenhouse gas emissions are accelerating, and symptoms of climate disruption are ubiquitous.[8] There is disturbing news of serious emissions of carbon dioxide from warming soils and possibly accelerating releases[9] of methane—another powerful greenhouse gas—in the vegetation carbon "sink" that helps to mitigate warming. Indeed, in July 2014 several large craters appeared in the tundra of northern Siberia that might have been caused by the massive release of methane arising from thawing permafrost.[10] Perhaps these are the first harbingers of the climate chaos to come. Methane seeps from warming ocean floors are potentially even more serious. As one climatologist eloquently put it, "If even a small fraction of Arctic sea floor carbon is released to the atmosphere, we're f'd."[11]

More worrying, our governments are refusing to act with the urgency that much of the scientific community thinks would be wise. The utter failure of the Rio+20 circus[12] in 2012 to achieve any meaningful international commitments on the environment made that crystal clear. Scientific concern is tightly tied to the energy situation. On the one hand, there is the need to transition rapidly away from the present rate of fossil-fuel use because of its role in climate disruption,[13] in a world in which few leaders seem to grasp either the need or the difficulty of this. On the other hand, there is also the prospect of rapidly diminishing supplies of high-grade oil and severe escalation of the costs of obtaining it,[14] when cheap oil now forms the key underpinning of civilization. In 2011 well over 80% of all human energy use was provided by fossil fuels. Humanity thus has two pressing reasons to hurry that transition, which without moving onto an emergency footing could take many decades.

Consider just the critical issue of food production. Without abundant, inexpensive oil, vast quantities of chemical fertilizers, pesticides, and fuels to run farm machinery, dry crops, and refrigerate and transport foods, food prices could easily soar beyond the reach of many.[15] Further down the road loom what might be even greater problems. One is the potential depletion of the easily accessible phosphate rock on which agriculture depends.[16] Another is the realization that many of the seeds stored frozen at the Svalbard Global Seed Vault[17] (on the Norwegian arctic island of Spitsbergen) will likely be useless for restoring the critical genetic diversity of crops. They consist of genetic "snapshots" of populations adapted to conditions that no longer exist[18] and have been largely selected for viability while frozen—hardly a critical characteristic for maintaining productivity as the world warms.

These biophysical problems are interacting with human governance systems, institutions, and civil societies that are no longer adequate to deal with them.[19] Typically, studies of coupled human and natural systems focus on reciprocating interactions and feedbacks between social systems and their biophysical environments. But a major challenge today for scholars in this area is to determine whether enough of such coupling remains between the two gigantic systems, biophysical and socioeconomic-cultural, or whether society is simply plunging ahead without reacting effectively to the deterioration of the biophysical environment. Thresholds for serious climate disruption are passing, toxification of Earth is proceeding apace and producing worrying symptoms, losses of vital biowealth are at a 65-million-year high with potentially extremely serious consequences for ecosystem services, the epidemiological environment is deteriorating, and a race is building to control water flows and extract the last high-quality resources, increasing the chances of ending civilization in environment-wrecking nuclear and other wars.

The social system has attempted to respond to this complex of problems. In the 1960s and building on much earlier work, scientists began assessing the consequences of an ever-growing human population and expanding consumption, overuse of pesticides, radioactive fallout, air and water pollution, and other environmental issues—and to recommend ameliorative steps. In the mid-1980s, biologists formed the discipline of conservation biology, with the explicit purpose of stemming the hemorrhage of biodiversity. Later in that decade, important reaction to the worsening environmental situation was the development of the Montreal Protocol[20] to preserve the vital stratospheric ozone layer. Around the same time, it dawned on the scientific community that climate disruption was going to be more immediate and dangerous than previously thought, but here attempts by Australia and the United States to take mitigating steps have been pathetic, although a few nations such as Japan and European Union members were moving forward in the 1990s. Overall, action to deal with other dimensions of the environmental dilemma has been utterly inadequate.

To see the growing disconnect, one only has to consider the attention paid in public discourse to the *relatively* trivial financial problems currently facing the socioeconomic-cultural systems of rich nations. Financial coverage in the media is massive compared to, say, the news that Earth's coral reefs are now likely beyond saving or that we might be passing a threshold beyond which heating the Earth past two degrees Celsius is inevitable. Or consider the complete failure of most social scientists to come to grips with the inability of civilization to develop mechanisms to deal with global environmental problems, or of the persistence of an economic system based on unrealistic academic models and the preposterous notion that growth can continue forever.

Whether mutually beneficial human/nature coupling can be restored in time is an open question. It is doubtless that action from the grassroots would be required, as well as new institutions and mechanisms for coordinating bottom-up and top-down efforts. We have strong evidence that giving women equal rights and opportunities everywhere (this has yet to happen in any nation), and providing every sexually active person access to modern contraception and backup abortion, would mostly solve the problem of fast population growth, and, perhaps, before the century's end, even set human numbers into the needed global pattern of gradual decline.[21] We know from World War II mobilizations that consumption patterns can be changed virtually overnight when urgency requires it and the political will exists.[22] There are many hopeful small-scale efforts to deal with important parts of the human predicament, such as the Natural Capital Project[23] to protect biodiversity and ecosystem services, deployment of sustainable energy systems in many countries, and work to unite academics and civil society in developing the necessary foresight intelligence and coordinating their activities to focus on the most critical issues, as in the Millennium Alliance for Humanity and the Biosphere (MAHB).[24] Other bottom-up efforts such as Occupy Wall Street,[25] the movement to solve the climate crisis by 350.org,[26] the movement to divest from the stocks of fossil-fuel companies,[27] and many other civil-society groups are gaining some traction. But time is short, and in our view the decoupling is winning.

So while scientists talk of "coupled human and natural systems,"[28] the evidence suggests that the systems are actually decoupling and that the human component is no longer capable of reacting adaptively to ever more rapid changes in the natural system. The signs of collapse, especially diminishing marginal returns to complexity,[29] are everywhere. The challenge confronting *Homo sapiens* is, without hyperbole, gigantic. To adapt, humanity must rapidly revise its systems for mobilizing and distributing energy and end its heavy dependence on burning fossil fuels. The problems and pitfalls of doing this are generally underrated,[30] and even China (unlike Australia and many other nations), which is struggling to transition from coal, is a decade or more from even halting the rise in greenhouse gas emissions from its burning.[31] At last, however, the United States is beginning to get rid of coal burning—still possibly too little and too late. In contrast, the latest pearl of energy wisdom from Tony Abbott is that coal is "good for humanity."[32] The mind boggles.

The time within which a global transition must be accomplished is debatable, but the scientific consensus is overwhelmingly that it is short—with some distinguished climatologists pressing for a World War II–type mobilization.[33] Society must simultaneously provide critical irrigation water to agriculture and attempt to rebuild its deteriorated water/waste-handling infrastructures, while it continues to strive to supply clean water and sanitation to the billions of people who still lack them. If leading climatologists are correct[34]—and it is becoming increasingly apparent that even their cautious predictions are, if anything, far too conservative—it is likely that Earth's precipitation patterns will be changing continually for a millennium or more. Among other things, this implies that dams, pipelines, and canals will need to be rethought, redesigned for flexibility and resilience, and rebuilt. Replacing decaying water pipes and sewers in many cities is necessary but insufficient—it will not, for instance, solve the problem of watering vast areas of crops as they desiccate in newly forming drought regions. It also implies massive investments in agricultural research to develop strains of grains and other crops that are resistant to heat, drought, and flood, and to develop technologies to deal with the considerable increase in pest problems and dangers of toxics that will accompany warming and population growth. Of course, in the background are the colossal and extremely expensive problems of restructuring humanity's systems for mobilizing energy.

The difficulties of maintaining an environmentally safe energy supply will carry over to many other areas. Transporting (or desalinating) water—an enormous problem in Australia and the southwestern United States—requires energy, and lots of it. Clearing and preparing new cropland in attempts to maintain food production requires energy—think tractors, harvesters, plows, pickers, cold storage, and transport. So does winning metals and rare earth elements in a world where depletion of rich ores is well advanced for most.[35] More and more energy will be mobilized in attempts to maintain or increase human well-being, while that process of mobilization will most likely tend to *reduce* well-being.

Siloing—isolating information and expertise in separate disciplines—is the major feature of the culture gap and one prime cause of the public's relative lack of concern over crucial environmental problems. Consider some ominous trends that are largely unknown or unappreciated by most people, rarely dealt with in education systems, including, outside of specialized courses, at the university level, and poorly covered in the media. The concentrations of greenhouse gases in the atmosphere have continued to increase because of human activities, especially over the past half century; the feedbacks recently discovered in the climate system have mostly been positive (heating itself produces even more heating); and ever more signs of climate disruption have appeared, from thinning polar ice to a rising incidence of extreme weather events.[36]

Extinctions of populations and species, already occurring at a rate many times baseline—that is, the rate during periods outside of the five "mass extinctions" over geological time[37]—make it ever clearer that humanity is causing a sixth mass extinction episode. Human impacts on the biosphere are currently so large that this is now an era commonly referred to as the Anthropocene.[38] Today almost a billion people have too little food,[39] more than when Green Revolution technology was first deployed. About 2 billion people have micronutrient deficiencies that can seriously hinder capacity. For example, vitamin A deficiency is common especially among the poor; it causes visual problems and even blindness in children and makes them more susceptible to serious childhood diseases. Iron deficiency is the most widespread, affecting some 2 billion people in both rich and poor nations; the anemia that ensues reduces the work capacity of entire communities. These and other deficiencies tend to reduce the ability of people's immune systems to defend them. The number of immune-compromised individuals is thus also at record levels, furthering the decay of our epidemiological environment.[40] Along with the other epidemiological consequences of the population explosion, such as the spread of tropical diseases toward the poles facilitated by population-related increases in greenhouse gases, the emergence of deadly pandemics is becoming ever more likely. The 2014 Ebola epidemic could be an omen of much worse to come. By early November, cases in West Africa had passed 10,000, and about half of the infected people had died. There were scattered cases elsewhere and no end in sight. We hope the virus will not evolve into a more easily transmitted form, which could generate a global concern. The only thing certain is that civilization was not prepared for the deterioration of its epidemiological environment, closely tied to overpopulation, despite numerous warnings.[41]

The Himalayan "water tower"—the ice and snow of the Himalayas and the Tibetan Plateau—is melting, although there is variability and uncertainty about the exact rates. The melting is part of a general loss of Earth's ice cover on both land and sea, which among other things will affect patterns of climate disruption, sea-level rise, and seismic events in ways that are difficult to predict. The loss of the water tower will likely first cause a complex pattern of flooding and then drying of some of the rivers that supply agricultural water to much of southern and eastern Asia[42]—this is a region that 1.6 to 2 billion people call home, with more being added all the time. Where do you think those billions will try to escape to when the worst crises hit? Australia and the United States are certainly targets, with Australia's hard line against refugees arriving by sea[43] perhaps only a hint of the human deluge to come in the near future. Worse, rising temperatures further threaten the food supply of those Asian nations,[44] three of which have rather frightening stocks of nuclear weapons. This is of major importance given that it has recently been shown that a "small" nuclear weapons exchange, say between India and Pakistan, could end global civilization as we know it.[45] The vexed problems of international migration are too complex to analyze here—it is enough to say that if we mitigate the major environmental problems, we will also reduce the pressures on people to move.

This brings us to our last point—the fallacy that wealth generates environmental capacity. There is a myth among many of the socially conscious that is used, whether knowingly or not, to justify the accumulation of wealth and rationalize consumption. In some ways, it is the primary underpinning of our societies' addiction to economic growth, and the conscience alleviator of rich nations. The concept is known as the "environmental Kuznets curve hypothesis," or the "theory of ecological modernization." It was named after Simon Kuznets, a Russian-American economist from Harvard University who won the 1971 Nobel Memorial Prize in Economic Sciences for his work on economic growth. He originally developed it for trends in economic inequality in the course of development. Basically, the concept argues that a society's environmental performance—that is, how environmentally sustainable it is—follows a U-shaped relationship with per capita wealth, with more sustainable actions kicking in past some threshold of individual wealth. The concept can be visualized thus: being poor is environmentally bad, because developing societies tend to make a lot of mess as they grow; however, as society gains an educated middle class, has access to (money to buy or manufacture) cleaner technologies, and acquires an environmental mass conscience, increasing per capita wealth leads to a *decline* in environmental damage.

Admittedly, the original concept was meant more as an instantaneous measure of a society's environmental performance rather than as a longer-term measure of cumulative environmental damage. However, the state of a nation's environment is not an instantaneous outcome of its current society's capacity or values; as we have described in this book, it is a long-term accumulation of multiple, synergistic environmental damages that eventually deplete biowealth to the point where they compromise human prosperity and well-being. The Kuznets curve must therefore be considered on the cumulative axis.

Regardless, the evidence for the environmental Kuznets curve is equivocal at best, or completely wrong at worst. Indeed, a recent study ranking the world's worst environmental performers by country indicated not only that the United States and Australia were among the top ten worst countries (the United States was second only to Brazil, and Australia was ninth worst overall) in terms of their total (cumulative) contributions to endangered species, deforestation, land degradation, fertilizer use, fisheries exploitation, greenhouse gas emissions, and water pollution,[46] but that there was also no evidence for a Kuznets-like relationship between a country's environmental ranking and its per capita wealth. So much for justifying the rabid accumulation of wealth that typifies growthmania with the goal of environmental improvement. Scaling ourselves back is therefore the only way forward.

11 Save This House

Save this house. Party's gotta end. The welcome mat's worn out, and the roof will never mend. The furniture's on fire, this house is a disgrace. Someone change the locks before we trash this place.

"Save This House" (1990) by Canadian folk rock band Spirit of the West[1]

The scale of the problems facing humanity is both daunting and, certainly in the United States and Australia, almost completely underappreciated by their citizens. This is due to a combination of interrelated factors that we have already addressed, including broken or at least highly inefficient education systems, ignorant politicians and other "decision makers," corrupt corporations, misplaced greed, and a widespread belief in the supernatural that the faithful assume will intervene to "save us" (from ourselves?). We do not wish to be mistaken for Pollyannas—our bias is actually in the other direction. We fear society will not get its act together in time and will instead count on praying to deliver us. We fear the Endarkenment will deepen, and the world will be forced to live in a faith-based rather than evidence-based Dismal Age.

Starting before agriculture with the control of fire, people have been severely altering their environments. Going way back—no one is sure how far, but in some groups more than 10,000 years—violence including warfare has been a feature of human behavior. But what evolution did not "count on," for it can have no foresight, was people learning to use their large social brains to extend their physical capabilities through the invention of many increasingly potent technologies. There was a limit to how many people an individual could kill with a stone club; however, there is essentially no limit to the number at risk from a ruler with the power to launch missiles tipped with thermonuclear weapons or our ability to design biological weapons against which there is no defense. Plows pulled by animals could and did change landscapes, but cheap oil, tractors, large-scale irrigation, and synthetic organic chemicals have changed the environmental systems of the entire planet. We are the uncontested champions at increasing the Earth's carrying capacity for ourselves, but only to a rapidly approaching limit.

The technology explosion now provides humanity with a whole new problem. That we have never learned to live happily and peacefully in large groups has led to gigantic suffering in the past, be it in the wars and collapses of ancient empires, crusades, *autos-da-fé*, or holocausts. But now technology, in combination with an insane growthmanic economy and "ethical" systems rooted in ancient delusions and myths, threatens for the first time to bring down a *global* civilization.[2]

A global collapse could involve anything from the spreading effects of nuclear or biological terrorism or a "small" nuclear war, whose ecological effects could quickly end modern societies as we know them,[3] to a more gradual breakdown, as escalating economic inequity, famines, epidemics, and resource shortages cause a disintegration of central control within nations and breakdowns in trade at all levels. Joseph Tainter, who wrote a classic book on collapse[4] and was concerned with subglobal units, also emphasized that collapse was likely when people lose confidence in their governance systems. But today there is no real global system to engender such confidence, and the very heterogeneity of global societies makes any judgment of who will suffer what consequences in a collapse highly uncertain. Will it be the poor who suffer the most, being already on the edge, or the rich with farther to fall?

But a downfall is not preordained. Many things *could* be done that would reduce the chances of society undergoing some form of collapse. Unhappily, none of them is being done on the scale required to make much of a dent in the problems. For example, there is a lot of talk globally about alternative energy sources—switching from burning fossil fuels as humanity's main way of mobilizing energy and moving to some form of solar/wind/geothermal/wave energy and/or increasing the efficiency with which we consume energy. But as we have discussed, fossil-fuel consumption is increasing, and, indeed, so has the amount of CO_2 in the atmosphere. In the United States, there is a growing unemployment problem that can be traced in part to the thieves of Wall Street, but also to the lack of foresight of our society, especially a failure to reduce economic inequities and shorten the working week. The threat to civilization of climate disruption could be reduced by removing subsidies for fossil fuels and transferring them to various forms of renewable and non-fossil-fuel energy sources, especially solar but possibly including adoption of new energy-dense fission technology based on the full recycling of spent nuclear fuel (what we currently call "nuclear waste").[5] In doing so, large numbers of people could be put to work building, installing, and maintaining these technological systems. Individuals could assist by refusing to invest in fossil-fuel enterprises or actively boycotting businesses that do. There is some motion in this direction, but it is much too slight. Transforming the United States and Australia into nations designed for people rather than automobiles will be required over the next half-century or so (the original mistake of suburbanization stretched over roughly a century, but of course it was no accident and reaped monumental profits for its perpetrators in the automobile, tire, gasoline, road-building, etc., industries). That, along with other required changes—such as reworking the crumbling US water-handling infrastructure and the Australian farming and water-management systems—would generate a lot of jobs during the decades required to redesign the economic system.

So let's take a closer look at potential solutions to our predicament. Today the socioeconomic-political "complex adaptive system" (CAS) of human civilization has reached a scale at which it has begun to influence considerably that other all-important, large-scale CAS—the biosphere. Most people who pay attention are now at least aware of a complex nexus of population-resource-environment-equity problems that have been thoroughly explored by the scientific community.[6] In this chapter, rather than trying to summarize the entire predicament or reiterate solutions already put forth (such as getting the money out of politics), we make, and sometimes repeat for emphasis, points that remain generally unappreciated but must be widely understood if solutions are to be found and adopted. We then explore some of the manifold changes we believe will be essential components of the solution to the human predicament, for it is abundantly clear that there is no single "magic bullet" that will save civilization. Complex problems invariably require complex solutions. Likewise, we avoid spelling out the obvious "What can I do?" actions of individuals. While eating less meat, recycling, installing solar panels, buying nontoxic cleaning products, and divesting from fossil-fuel enterprises are certainly good things to do, these actions by themselves will not bring about the fundamental changes to civil society and our economic systems. We instead argue that the strongest action an individual can take is to overwhelm our corrupt and special-interest political systems by voting for those that set long-term sustainability of the planet in their sights—by addressing the problems and following the actions we outline next—instead of capitulating to the short-term interests of their most influential and cashed-up lobbyists. Political action from the bottom up seems to hold the most promise for solving the environmental crisis, but what actions should people press for? Here's a summary; many of the solutions will appear "impractical," but they are in fact a hell of a lot more practical than inaction.

THE PROBLEMS AND SOLUTIONS IN BRIEF

Population Growth Is a Major Contributor to Environmental Problems

One of the key threats that civilization faces today is climate disruption, and even though it is almost never mentioned in the mass media, the flow of greenhouse gases into the atmosphere from driving cars, transporting goods, deforestation, eating beef, and heating and lighting homes and businesses obviously is tightly tied to the number of people on the planet.[7] Population size multiplies with per capita consumption to produce a level of *aggregate* consumption that today is unsupportable in the long term. Both Australia and the United States have some of the highest per capita emissions and resource consumption rates.[8] It makes no sense to say that population size or consumption per person is more responsible for the human predicament—the two components are inseparable.

Solutions: Give women equal rights and opportunities, and make modern contraception and backup abortion available to all sexually active individuals. These won't be fast solutions, but their global-scale implementation is way overdue.

Climate Disruption Is Not Necessarily the Most Serious Environmental Threat

This notion is not appreciated widely or even a major societal concern. For example, a small analysis from the BBC in the United Kingdom demonstrates the disparity in the public domain: over three months in 2009, UK newspapers mentioned "biodiversity" only 115 times compared to 1,382 times for "climate change."[9]

Up until very recently,* there was no biodiversity equivalent of the Intergovernmental Panel on Climate Change, and we still have little capacity or idea how to account for the trillions of dollars' worth of biowealth supplied every year to us free of charge. Neither do we have anything at all equivalent to the Kyoto Protocol aimed at biodiversity preservation. Yet conservation biologists have for decades demonstrated how human disease prevalence, reduction in pollination, increasing floods, reduced freshwater availability, carbon emissions, loss of fish supplies, and weed establishment and spread are all exacerbated by biodiversity loss. We are not talking only of the loss of the diversity of species. Currently the loss of the diversity of populations, local subsets of the creatures that are lumped together in classification and called the same "species," is much more important. Remember, populations are what deliver the critical natural ecosystem services on which all of humanity depends. Climate disruption, as serious and potentially apocalyptic as it is, can be viewed as just another stressor in a system already stressed to its limits—one weapon used to pummel populations and species to extinction among many others in the armory.

Of course, the lack of interest might not be as bleak as indicated by mentions in newspapers; we contend that the science underpinning our assessment of biowealth loss is fairly well accepted by people who care to look into these things and that the evidence spans the gamut of biological diversity and ecosystems. In short, it is much less controversial a topic than climate disruption, so it attracts a lot less vitriol and spawns fewer polemics. That said, it is a self-destructive environmental

*The new Intergovernmental Platform on Biodiversity and Ecosystem Services (IPBES) is just getting some momentum (www.ipbes.net).

One of the rarest birds in the world (approximately 50 remaining in the wild and 250 in captivity), the orange-bellied parrot (*Neophema chrysogaster*) is the epitome of imminent extinctions unfolding around the planet. Volunteers are working hard to improve its breeding conditions by putting up nest boxes and releasing captive individuals. Photo by Paul R. Ehrlich from southwestern Tasmania.

ambivalence that will eventually come to bite humanity on the bum in the most serious of ways. We truly fear that we are not far off from more major world conflicts (after the oil war in Iraq[10]) over the dwindling pool of high-grade resources (farmland, clean freshwater, concentrated mineral stocks, and biodiversity) we are so effectively depleting.

Solutions: Press decision-makers to heed the warnings on loss of biowealth, toxification, pandemics, and the like. Work to remove the media from corporate control, to improve environmental education, and to replace gradually the intricacies of sports coverage with the intricacies of civilizational survival.

Economic Growth among the Already Rich Hurts Everyone

Even with a substantial portion of Earth's human population living in poverty, more than one planet would be needed to sustain today's population indefinitely. Estimates vary, as does the definition of "poor," but if we take just the number of malnourished "hungry" people on the planet, it is today nearly 1 billion. If we take the number of people living on less than US$2 per day and those who are micro-nutrient malnourished, the number of poor could be as high as 3 billion.[11] Thus, the more consumption by the rich grows, the less there will ultimately be for others, and eventually for the rich themselves. Wealth redistribution will be necessary regardless of how it comes. Hopefully we can avoid a huge-scale version of the French Revolution's "Terror" by choosing to modify our grossly out-of-kilter economic systems, but we fear heads will roll nonetheless. As world-renowned economist Joseph Stiglitz writes:

> The top 1% have the best houses, the best educations, the best doctors, and the best lifestyles, but there is one thing that money doesn't seem to have bought: an understanding that their fate is bound up with how the other 99% live. Throughout history, this is something that the top 1% eventually do learn. Too late.[12]

Simple mathematics shows that a "long" history of exponential growth (in this case, just around 250 years out of millions) does not imply that a long future of such growth is possible.[13] Such a basic misunderstanding not only demonstrates a severely underdeveloped mathematical education in society (including in most universities); it shows that people in general place far too much faith in human technological capacity than is possibly justified.[14] Perhaps the most extreme techno-fix, colonizing other planets with the billions of extra humans our globe now has in excess, can be shown to be impossible by simple arithmetic. Suppose, for instance, that the ten largest airlines (those that fly the most passenger miles) could fly to Mars at the same rate they carry people around Earth, and that Mars was fully habitable and welcoming. If they did nothing else, they could carry about 5,000 people to Mars annually. If they could fly 100 times as fast (spaceship speed is about 62,000 miles per hour, or around 100,000 kilometers per hour), they could transport some half a million people a year. But the human population is growing by roughly 80 million a year. Furthermore, even if Earth and Mars were identically habitable, Mars would be as densely populated as Earth with less than 2 billion people,* because its surface is only about a quarter as big. Simple arithmetic. The romantic science-fiction dream of "offworlding" to save humanity from a dying planet is clearly ridiculous.

Blue Planet Prize–winning economist Herman Daly, a leading analyst of the steady state economy, summarized:

> Why . . . do we keep growth as the top national priority? . . . [B]ecause even though the benefits of further growth are now less than the costs, our decision-making elites have figured out how to keep the benefits for themselves while "sharing" the costs with the poor, the future, and other species. The elite-owned media, the corporate-funded think tanks, the kept economists of high academia, and the World Bank—all sing hymns to growth in perfect unison, and bamboozle average citizens.[15]

While neither growth in inputs to or outputs from the global economy can go on forever (or indeed for much longer), there is one area of growth that obviously could go on for a very long time—perhaps long enough to be considered "forever." That, of course, is growth in well-

*The surface area of Mars is about 144 million km^2, which is nearly the same as the total land surface area of Earth (148 million km^2). In this example we're therefore assuming the same proportion of land and sea in an Earth-like "habitable" Mars.

being for all people. Unhappily, although there is a tremendous need for such growth, humanity seems locked into an economic system designed to prevent it.[16]

The culture gap[17]—the extreme compartmentalization of education and knowledge that keeps most people from understanding the basics of how the world works—makes finding solutions to humanity's existential problems extremely difficult. Few grasp how fundamental this problem is. Without knowledge of topics such as how the climate system functions; the environmental nonlinearities associated with population growth; the necessity of biowealth for human health, wealth, and well-being; and diminishing marginal returns to complexity that can signal the approach of a collapse,[18] even many well-educated people are in a poor position to make sound political decisions.

Consider the situation in one central, supremely important, educational silo of economics. While some economists have done crucial work on how to deal with environmental problems,[19] many remain sadly ignorant about them. Consider a paper by economist David Lam about the population explosion, a crucial driver of environmental destruction, entitled "How the World Survived the Population Bomb: Lessons from 50 Years of Extraordinary Demographic History."[20] The title itself implies that the projected addition of some 2.5 billion people in the next forty years (equivalent to the entire population of Earth in 1950) is not part of that bomb. In fact, the nonlinearities that will inevitably accompany such growth will greatly expand its negative environmental impact. Unhappily, some economists act as if the first human being was Adam Smith and think in terms of decades about the situation of the human organism with billions of years of history behind it. Those years were riddled with intense planetary changes—some on time scales of decades or less and only during about 200,000 of those years was the human organism modern *Homo sapiens.* The Lam paper does not explain that more people are probably hungry today[21] than in the 1960s when the *Population Bomb* was written, that conservatively more than 300 million have starved or died of starvation-related disease since then, that today one child in four is too malnourished to grow properly,[22] and that the short-term success of the Green Revolution might have been bought at the price of long-term disaster. In India that revolution has been involved in roughly 100,000 farmer suicides.[23]

India further illustrates today the ways in which overpopulation contributes to massive misery and death. Half of India's population—more than 600 million people—lack sanitary facilities and are forced to defecate outdoors, and almost half of Indian children under the age of five have stunted growth. The nation's rivers and streams are open sewers, turds are ubiquitous, and dried fecal matter blows as a pervasive dust in the winds. Thus, even well-fed children are immersed in a miasma of pathogenic bacteria and must divert metabolic energy into fighting disease rather than to growth and brain development as would normally happen.[24] Population growth has overwhelmed otherwise inadequate attempts to provide more toilets, and even where they are provided, Hindu traditions militate against their use. Furthermore, childhood stunting is more prevalent in India than in much more poverty-stricken sub-Saharan Africa because the population density in India is roughly tenfold that in Africa. That stunning gap is sadly likely to close from the wrong direction because Africa's population is projected to double or more by 2050, while India's will grow by "only" about 30%.

The greatest oversight in mainstream economics articles such as Lam's lies both in ignoring the extreme seriousness of environmental problems and not realizing their dynamic connections with demography.[25] These articles normally do not even mention decay of ecosystem services; loss of natural capital, biodiversity, and genetic diversity of crops; planetary toxification; increasing probability of wars and pandemics; or nonlinearities, thresholds, feedbacks, synergies, or tipping points[26] in relation to population and economic growth.[27] Many economic works today[28] seem to share the implicit belief that there are no limits to growth. As distinguished economist Kenneth Boulding wrote in 1966:

> Anyone who believes that exponential growth can go on forever in a finite world is either a madman or an economist.[29]

Solutions: Change the focus of economics to get rid of growth as a major goal and transfer emphasis to distribution. Push for caps on personal wealth and a much more equitable distribution of natural resources among nations. Teach everyone that perpetual growth as imagined by some economists is biophysically impossible. Try to teach everyone elementary mathematics, and educate all of society to discard the myths of growthmania. Press to begin an examination, academically and through public discussion, of the pressing questions of how to organize a steady-state economic system, and revise the antiquated structure of education globally to close key parts of the culture gap. It is even time to examine the notion held by many economists that human well-being can grow forever. Obviously that would involve defining "well-being," and by any definition there are billions of people who could undergo enormous growth in that aspect if they acquired the physical, educational, and other standards enjoyed by most people in rich countries today. But once they caught up, could, say, 3.5% annual growth in well-being then continue forever? In a few hundred years, everyone would be a thousand times better off. What would that mean?

All These Factors Are Intertwined

The interactions among these problems are daunting.[30] Meeting human needs means increasing agricultural production some 70–100% by 2050 in response to population growth (unless huge steps are taken to limit it), rising demand in emerging economies for meat-rich diets, and increasing demand for biofuels, all of which will require a larger oil subsidy to agriculture. The enormous complexity of what needs to be done just to supply the food needs of today's population is daunting. What is crystal clear is that no single "magic bullet" approach will suffice even there, and the prospects of making 2.5 billion additional people food-secure by 2050 is extremely problematic.[31] The results will be more greenhouse gas emissions and land degradation and higher food prices. Increased emissions will alter precipitation and temperature patterns further, threatening farm productivity. Any half step forward in increasing agricultural productivity and/or technology could be offset therefore by two steps backward—a net *reduction* in food production and a restricted capacity to distribute what food is produced more equitably. Increasing both yields and total food production will be made more difficult by the environmental effects of moving to more intensive and extensive agriculture, including accelerated losses of biodiversity, soil erosion, and depletion of vital aquifers, as well as further toxification of those aquifers, surface freshwater, soils, oceans, and organisms. The increased probability of famines is not the only consequence of the population explosion; others include growing chances of vast epidemics, resource wars, battles against increasing flows of resource and environment refugees,[32] and deaths from violent climatic events.

Solutions: Besides changing the power structure of societies, striving for equity, and redoing education systems, human priorities need to be changed. Everyone should start paying much more attention to humanity's primary activity—growing and gathering food. Fueling people is much more important than fueling automobiles (or cattle).

Australia and the United States Play Different Roles in Both Problems and Solutions

Now let's look in more detail at some of the things that Australia and the United States might do about the human predicament. Despite their great cultural similarities and long friendship, the United States and Australia for geographic and historical reasons inevitably play roles in the world that are strikingly different. Resource richness and the First World War vaulted the United States into a position that, with the help of the Second World War (really, the second act of the First World War), allowed it to achieve near-global hegemony. The United States, with ideas rooted in notions of American exceptionalism, built an empire, most recently on a rather new basis. It did not continue in the twentieth century to emulate the earlier European model of continually occupying more territory as it had in the nineteenth century. Rather, it used its military and economic power to bend other nations to its will. That has been in no small part a project to control sources and flows of oil, the fuel upon which the modern world has been built. The United States originally did this by supporting dictatorial regimes in nations deemed to have strategic or security interest for the United States. More recently, it has switched to "spreading democracy"—this translates to supporting friendly elites in positions of power in poor countries and keeping their populations docile as consumers of US exports and exporters of raw materials needed by the empire.[33] In essence, the main policy of the American empire is now supporting a global elite seeking to dominate access to global resources. Alan Greenspan, chair of the US Federal Reserve System, let this slip in 2007 when he famously stated:

I am saddened that it is politically inconvenient to acknowledge what everyone knows: the Iraq war is largely about oil.[34]

Australia has been a close, if very junior, partner to the United States in its imperialistic enterprises, prominently displayed in 1966–67 Prime Minister Howard Holt's enthusiastic support[35]—complete with an "All the Way with LBJ" speech—of the gigantic and immoral US blunder in attacking South Vietnam. Australia saw and sees its American connection as the source of its military security. Under Prime Minister John Howard, it attempted to walk a narrow line between support of the aggressive Bush-Cheney doctrine and a more intelligent multilateral approach to international issues promoted by China, the other elephant in its room. Subsequent Australian governments have followed a similar course.

Australia has not suffered as much from hypocritical imperialism or blind exceptionalism as the United States has. If anything, Australia has been the victim of a long-term inferiority complex, downplaying its accomplishments and leadership potential. In Australia this complex is expressed socially as the Tall Poppy syndrome,* where people who are talented and succeed in life are often criticized for that success. We would hope that Australia could now seek to become a positive international example as Canada was in the 1960s under the brilliant leadership of Lester Pearson, rather than emulating under Tony Abbott the disgusting descent of Canada into an "environmental rogue state."[36] That, sadly, is what Canada has done under the rule of anti-science, anti-environment Prime Minister Stephen Harper.[37] Australia is at the front lines of a potential revolution in international governance—the replacement of US hegemony by a multilateral system in which, among others, China would be a major player. The United States, still pursuing a version of the Bush doctrine (mostly shorn of its pseudo-democratization rhetoric[38]), is clearly headed toward military confrontation to stem its declining influence in East Asia. Australia can

*The expression apparently originates from several Greek and Roman accounts of cryptic political messages manifested by cutting off the heads of poppies or wheat that exceeded the average height of the crop. The symbolic gesture was meant to advise autocrats to slay citizens of outstanding influence or ability who would otherwise challenge their authority. In other words, it removed the most likely competitors.

play along in this potentially catastrophic game, or it can move into a leadership position to create a new form of multilateralism, based in the severe threats to the security of China, the United States, Australia and the rest of the world from today's rapidly multiplying environmental problems.

Australia could reshape ideas of security, where it has a head start on the United States in (after some back-and-forthing) that it abjured nuclear weapons. Unlike the five original nuclear-weapons states (the United States, Britain, France, Russia, and China), Australia actually honored its commitments after signing the Treaty on the Non-Proliferation of Nuclear Weapons. The weapons nations, bound to further nuclear disarmament, have been remarkably reluctant to do so, and the United States and Russia still possess thousands of warheads, many apparently on high-alert status. Australia could take the lead in insisting that since the nuclear-winter studies[39] showed that no nation could win a large-scale nuclear war, and now we know not even a "small" one, all nations should back away from having nuclear missiles capable of being launched quickly. That, at least, would reduce the odds that billions might die by accident—hardly a zero-probability event since it nearly happened in 1995.[40] Australia could further push for enhanced international control over stocks of plutonium and highly enriched uranium to make it more difficult for terrorists to create and deploy nuclear weapons.

In our view, the threat of a nuclear weapon detonating in a major city is one of the most serious and underappreciated elements in the storm of environmental problems. One only need consider the fate of US civil liberties and the global consequences of the 9/11 attack, a conventional assault killing a few thousand people, to imagine what might happen globally if a few million were killed in Washington, Sydney, Moscow, or Beijing were they attacked. If Australia wants to be sure the Sydney Harbour Bridge and Opera House will be standing in a century, it could become a constant pain in the international bum, pressing for the dismantling of nuclear weapons and greatly increasing the security on those not dismantled. It should vigorously oppose new states joining the weapons club (North Korea) and equally push the United States and Russia to lead the way symbolically by securing and destroying their gigantic arsenals. It should also oppose new production of the essential ingredients of bomb building: plutonium and highly enriched uranium (Iran). It could expect some help from the United States on these projects, where many scientists and government officials share the nuclear terrorism concerns, where the size of the nuclear arsenal has been greatly diminished (although there is no sign of even an approach to the ideal warhead number of zero), and where the nation has been converting the fissile material in warheads purchased from Russia into electricity.[41] But the bottom line is that Australia—with its excellent science, clear ability to arm itself with nukes, but with a public long in opposition to weapons of mass destruction—should step out from beneath America's deterrence-woven skirts and start leading the globe toward total nuclear disarmament.[42] It could begin with repeatedly pointing out that detonation of only a small number of nuclear warheads, perhaps fewer than a dozen, would end the United States or China as a functional entity, fewer than that for Russia, and six for Australia. Using many thousands would simply bring down the curtain on civilization through nuclear-winter effects.[43]

But aside from such elements of traditional (military) security, Australia, with arguably the best group of environmental scientists per capita in the world, could be leading the way in refurbishing and reorienting the international system to deal with the other threats facing global society, such as climate disruption, loss of biowealth, and toxification, all threatening humanity's feeding base and escalating the odds of epidemics eroding human health. It could shift from being an ally of the United States in the war against the environment to being a pioneer in seeking ways to increase human happiness by reducing its own population size and per capita consumption of resources while freeing up more for the needy poor. It could focus its technological expertise and ethical "soft power" (that generated by its non-economic culture, including its foreign-policy positions) on issues of access to mineral resources and the technical Red Queen dilemma—the need to run faster technologically to maintain a decent standard of living in the face of continuing degradation of those resources.[44]

Above all, Australia could build on its own experience in debating population size to advance the issue globally—in this regard the United States is one of the most backward nations in the world. Many Australians realize how overpopulated the nation is, especially with the prospects for regional climate catastrophe as telegraphed by the boiling 2011–12 "Murdoch Summer"—so named after the man who has done most to promote climate-disruption denial. Australia's soft power could also be applied to the building disaster of transnational migration, especially the issue of climate refugees—people fleeing vulnerable lands and trying to enter the rich nations primarily responsible for climate disruption. Australia might even start an international discussion on the ethics of borders—on how an overpopulated globe can deal safely and ethically with the uneven distribution and serious depletion of resources. Australia, the world's twelfth most powerful economic force, still might plant the seeds of the sorely needed global reorganization that its record of domination would prevent the United States from doing. On the other hand, the nation must rid itself of the Abbott curse as rapidly as possible, or likely lose much of its influence and drop substantially in soft-power rankings.[45]

Australia must also promote broad discussions of resource issues focusing on the transition to emissions-free technologies; Australia is potentially a technological leader in both renewable and possibly nuclear power[46] (not weapons) technologies. In place of the country's current self-destructive push on coal for its own electricity generation and export,[47] a radical shift toward decarbonization is essential. There are knowledgeable politicians in Australia whom one imagines could lead such an effort, such as former foreign minister and previous long-term premier of New South Wales, Bob Carr, or Senator Christine Milne, former leader of the Greens, or her predecessor Bob Brown. Otherwise, there are unfortunately slim-pickings in this area among most of the political scene in Australia.

Internally in both nations, a barrier to sound policy derives from the oppressive weight of what many imagine to be ancient religious beliefs, but are actually modern religious politics. This was recently highlighted in the United States by the *Hobby Lobby* decision putting religious policy ahead of women's health, and by the incompetent, ignorance-based governance of Tony Abbott, who repeatedly refers to god as backing his environmentally malign policies.[48] But even here the United States could still learn from Australia. In the United States, politicians are required to name-check a god whether or not they believe it its existence, whereas in Australia there are considerably more constraints on claiming the political support of an imaginary supernatural entity; and Julia Gillard, an open nonbeliever, unmarried with a partner and childless, was recently prime minister. In any case, a transition toward evidence-based instead of faith-based policies in both nations could increase the odds of avoiding a global collapse, and Australia is better positioned to lead the way despite the retrograde Abbott government. The United States could certainly learn from its less, but increasingly, Christ-bitten ally.

LEARNING FROM EACH OTHER

Business as usual is an increasingly dangerous option, virtually certain to lead to a collapse of civilization that could feature such things as a breakdown of order within political units, conventional or nuclear wars over resources, a cessation of international trade, pandemics or famines that kill billions, and a global temperature increase of up to 5–10 degrees Celsius. If some form of a collapse is to be avoided, determined action on many fronts will be required, mobilized with the sense of urgency that swept the United States at the start of the Second World War. A sample of the steps we think should be taken, outlined in the summaries, follows, starting with two important ones that target the principal drivers of environmental deterioration: (1) overpopulation and (2) overconsumption by the rich. In this regard, the United States and Australia could pick among the best of each other's historical policies and approaches, while avoiding the disasters of bad decisions that have hampered progress in each. Then we point out areas where symptoms need to be addressed and, finally, note the most important solution: changing fundamentals of human behavior. Of course, our list is far from comprehensive, but as stated earlier, it represents what we think are the most important gaps not being considered fully by either nation today. Throughout one should keep in mind that there can be a substantial difference between individual values and those promoted by institutions.

Empower Women

Neither Australia nor the United States can take pride in their records in this area, although Australia is making some strides toward reducing discrimination and violence against women.[49] In Australia and the United States, a wide range of birth-control options is available to women with the knowledge and means to access them. Nonetheless, in both countries, more than 50% of pregnancies are unintended, women often do not confer with physicians about contraception, and physicians are often not helpful when they do.[50] Australian abortion law is basically a mishmash of criminal codes controlled at the state level,[51] although overall the situation seems superior to that in the United States, where the Republican war on women, backed by a majority of male religious zealots on the Supreme Court, has been gradually pushing back the advances made by the 1973 *Roe v. Wade* Court decision,[52] basically decriminalizing abortion nationally until late in pregnancy. Australia should learn from the past national progress in the United States and hope, along with all ethical people, that the crazies attacking women—including Abbott and his sexist henchmen[53]—represent a sad and temporary aberration.

Correcting those inequities must be a top priority, both in the name of simple justice and because, in general, liberated women have fewer children and make more responsible decisions about household consumption.[54] The size of the unmet demand for contraception is somewhat controversial,[55] but it seems likely that each year as many as 50 million unwanted pregnancies and tens of millions of abortions (many dangerous) could be avoided with an investment of some $4 billion, well under the cost of one aircraft carrier. The investment would be retrieved by reduced costs of poor health and loss of life. Additionally, families with fewer children can invest more resources in education and health, setting up future generations who would be better able to understand the complexities of population, environment, and development. Empowering women and giving them equal rights and making modern contraception and safe backup abortion available to all sexually active individuals are perhaps the most important elements in the solution of the human predicament. Combined, they alone might move humanity toward desperately needed shrinkage of the global population to a more sustainable number. The human enterprise is much too large for safety; it must be rescaled, and nations such as Australia and the United States should be reinforcing each other and leading the charge.

Reduce Overconsumption

Basically our super-consuming societies, epitomized by US and Australian lifestyles, must be brought to an end. But while solutions to overpopulation have been thoroughly examined by scholars, the issues of overconsumption have barely been discussed.[56] How to deal successfully, efficiently, and humanely with the necessary redistribution from rich to poor has not been carefully considered. What is the best way to control the scale of total consumption? Pioneering economist Herman Daly long ago suggested a program of depletion quotas that would limit the rate at which natural resources would be "severed," along with birth quotas—basically a scheme for purchasing the right to have more children than a standard allotment in order to reach a sustainable population size.[57]

Reducing the population and consumption drivers is essential, but they present vastly different challenges. It is not possible to make major changes in population size humanely (or sensibly), except on a timescale of many decades.[58] It is crucial that society start on this critical task now. It is not at all clear how long humanity can remain in its present state of vast overshoot before collapse is inevitable. But society has knowledge to reduce consumption, and the trajectory could be modified rather rapidly. Working out how to manage the vexed problem of redistribution, however, will be an unprecedented challenge. Still, societies have shown that they can change their patterns of consumption virtually overnight given the right incentives.[59] A major challenge for a revised discipline of economics would be working out the most feasible, efficient, and equitable ways of radically changing consumption patterns within a few decades so that the poor can consume more while the rich consume less, such that aggregate consumption is substantially reduced. Of course, if behavioral and political scientists could find a way to eliminate money from politics, true reforms that threaten the plutocrats could be taken up.

It is unclear which of the two nations here are better aligned to achieve such a consumption revolution, considering that both Australians and Americans are superlative wasters and overconsumers. The inevitable policy shifts that will arise following water shortages and mega-wildfires in both countries could see smarter use of at least precious water resources, and advances in electricity production and consumption technology—especially in the technologically more advanced United States—could start to address both nations' deplorable histories of inefficiency and waste.

Treating the Symptoms

Curbing overpopulation and overconsumption of the rich will be critical to prevent escalating catastrophes in the mid-range future, but preparation for deploying solutions to the disasters already entrained is also important. Symptoms of global overshoot are already becoming obvious. As society deals with the drivers, it must also take steps now to see that the symptoms—climate altering faster than expected, biowealth disappearing, weird changes in animal and human development, failure to feed everyone—do not bring down civilization.

Much attention has been given to the tasks of rapidly reducing the flux of greenhouse gases into the atmosphere and taking steps to ameliorate the impacts that are already inevitable,[60] but there has been far too little action so far. For instance, the solution of reengineering the world's water-handling infrastructure for flexibility so that water can be easily redirected to crops in a world of continually changing precipitation patterns has hardly been discussed. Society also needs to reorganize defenses against epidemics. In addition to curbing the population driver and feeding malnourished people, much more attention should be given to measures for isolating the ill, developing new vaccines and antibiotics, banning the use of antibiotics in animal feed and using them in medicine as scalpels rather than shotguns, stockpiling medical supplies including isolation gear, planning how to handle huge numbers of patients, and so on.

The scale of the required treatment of symptoms can be seen just by examining humanity's most important activity: feeding itself. A recent comprehensive analysis of food production and the environmental damage caused by agriculture[61] recommends halting the expansion of farmland in the tropics, compensating by increasing production on land already cultivated, while reducing unsustainable uses of nutrients, water, and agricultural chemicals, and shifting diets (including eating a hell of a lot less meat[62]) while decreasing waste. This is a gigantic challenge, especially considering the inadequate global governance and economic systems that encourage poverty and maldistribution of food. But the study does point in a direction that, if taken, could go a long way toward solving the human food problem.

Of course, following the political correctness that strangles sensible discussion of the global predicament, no mention is made in that article of dealing with the key driver and humanely reversing population growth as a major contributor to a possible solution. Similar comments could be made about a set of partial solutions to human health and climate change through abating the emission of black carbon and ozone precursors into the troposphere,[63] an area where real action is more likely. While Australia has recently regressed in the area of leadership in climate-change mitigation, at least the current US administration has been mostly talking the right talk (the major exception being its surface enthusiasm for drilling for gas and oil) while bought-and-paid-for representatives in Congress effectively block action to preserve the nation's future. Here again the ability of the super-rich to control the media and who gets an electoral edge looms large. The United States could learn from Australia's former and admittedly inadequate carbon-pricing scheme to engender real change in global emissions reductions, and Australia could essentially decarbonize in only a few decades by embracing at least some nuclear power generation, providing that satisfactory solutions to the waste issues can be resolved.[64]

Fix the Energy Problem

The social, political, and economic challenges to achieving a global-scale environmental turnaround are massive.[65] But time is of the essence, for how does one initiate such change fast enough and at sufficient scale in corporatocracies while still upholding the remains of democratic and liberal freedoms?[66] Can technological shortcuts assist? As we have discussed at length, energy provision will be the key to making the necessary changes while feeding our growing populations before they have a chance to decline to a sustainable size. Growing food, transporting goods, disposing of waste, providing clean water, and furnishing heating, lighting, and air-conditioning, while simultaneously reducing greenhouse gas emissions and preventing the loss of natural capital—all these will require oodles and oodles of energy.[67] For starters, both Australia and the United States have enormous potential for mobilizing solar energy. According to analyses by John and Mary Ellen Harte,[68] for instance, an area two and a half times the US military's Yuma and Dugway weapons-testing reserves (over 3,000 square kilometers, or more than 1,200 square miles) in Arizona and Utah covered with panels could meet a sizable portion of US electricity demand in 2030, including that for electric vehicles. Australia has even more abundant land now considered "waste" and a much smaller projected demand. South Australia has already been moving toward meaningful solar and wind power penetration, starting under the leadership of Premier Mike Rann, although these renewable sources alone will not meet even that small state's total demand profile.[69] Solar capacity in the United States quadrupled between 2010 and 2014, a good sign, although it still is under 2% of the US energy mix.

There would be many associated costs with such a transition (largely away from fossil fuels), but Australia and the United States could lead the way. Other countries with less abundant, unoccupied sunny land might opt to supplement solar with, for instance, new generations of nuclear fission power,[70] although both the United States and Australia might benefit more quickly and completely from a higher penetration (in the case of the United States) or at least *some* penetration (Australia) of nuclear power electricity generation,[71] even though economic and "not in my backyard" problems would make it challenging. Or they might decide to cover all suitable roofs with solar photovoltaic panels, a direction in which many countries are already heading (and a system that reduces problems of energy distribution). Placing such photovoltaics on all suitable roofs in the United States in, say, a period of a decade could help ameliorate unemployment problems both through the need for installers and increased manpower needed for manufacturing the needed equipment. Depending on a series of factors such as timescale, the number of jobs created could easily exceed 100,000. One must remember, however, that such a move would also cause job losses in conventional energy mobilization, and net employment impact (depending on many factors and how the numbers are done) could be close to zero or even negative. The main benefits of getting away from fossil fuels would be environmental. The unemployment problem would be better dealt with by steps such as reducing birth rates and number of hours worked per week.

We need to put all energy options on the table, including nuclear fission technology,[72] and speed the deployment of various renewable energy technologies that are now under way. Yes, there will be costs to both nuclear and renewables, but as A. M. Weinberg wrote:

> Technological fixes have unforeseen and deleterious side effects—but so do social fixes, especially revolutions.[73]

Revision of Governance Systems

The big governance issues today have both international and intranational components. On the international front, the major *recognized* issue is whether democratic or authoritarian states will become the norm. In the United States, Australia, and many other putatively democratic nations, governance is more and more being taken over by the rich, the corporations they own, and their agents, such as the International Monetary Fund and the World Bank. New priorities need to be established in which, for example, in the United States, taxpayer dollars are not subsidizing the already incredibly rich oil companies and coal barons. Taxpayers should not be supporting an extraordinarily bloated military establishment aimed to a large degree at stealing other nations' resources, making enemies, developing high-tech weapons systems that are never used to subsidize Silicon Valley, and funneling profits to unscrupulous industries and corporate paramilitary establishments. This suggests that one place where Australia might learn from US experience is in examining its election finance laws. In the United States, the catastrophic *Citizens United* Supreme Court decision, as well as the *McCutcheon* decision, added to the intrinsic political power of money and might have crippled most efforts in that nation to take steps to avoid a collapse. Australia could lead the way toward a badly needed universal program of government financing of elections, one step toward getting the money out of politics. Another step would be creating legislation to prevent too much power in the media to be held by a few power-hungry plutocrats. Similar steps to support diversity in the mainstream media would greatly improve the situation in the United States, and it could move forward in its governance by mandating voting as does Australia and by getting serious about the changes to the Constitution required to remove "personhood" from corporations that can have neither empathy nor ethics. Of course, that is more readily said than done—a corporate plutocracy is increasingly in control of the United States and now threatens Australia's long and successful social governance traditions. We doubt the plutocrats who pull the strings will give up their power so easily.

The nation-state system's developing numerical giants, the BRICS (Brazil, Russia, India, China, South Africa) and a few other countries, are racing to embrace unsustainable development and use newfound economic and military power to exploit the poor of the world in a new colonialism designed to protect and enhance their own consumption.[74] China's purchase of farmland in Africa is an outstanding example of an attempt to assure future food and fuel supplies for itself while displacing local farm communities. Even in Australia—a country renowned for its poor agricultural land—foreign ownership of farmland is increasing,[75] including investors from China, Japan, Brazil, United Kingdom, Qatar, Sweden, Switzerland, and even the United States! The land-grabbing Olympic Games have well and truly begun, so it is essential that some form of leadership in regulating global land-grabbing take place. It is doubtful that the military and colonial past of the United States will endear other nations to its leadership, in which case Australia might be ideally poised to take those reins.

The main hope for diverting society from its headlong rush toward some form of global collapse of civilization might well rest with bottom-up pressure from civil society; an unprecedented global social movement, motivated by absolute necessity and fear of collapse, might be generated while simultaneously and partially healing the culture gap. There are already myriad efforts, including the Occupy movements, the Arab Spring and Extreme Citizen Science,[76] to recruit ordinary people to reform their societies, gather and analyze scientific data relative to our predicament, and press society to deal with climate disruption.[77] One bottom-up movement that has been gaining traction as we write this is pushing for universities and other actors in civil society to divest from stocks in fossil-fuel companies. This is basically a symbolic act, and the slowness of "great" universities to do it is indicative of how much they are in the thrall of corporate donors. Harvard has been at the bottom, refusing to act and explaining its behavior in an astoundingly meretricious statement by its president, Drew Faust. A Faustian bargain, indeed.

There are other good signs. Independent filmmakers are pitching in to try to save society,[78] and many new journals[79] and educational websites[80] are involved in the struggle. One new bottom-up organization is attempting to coordinate civil society efforts, building on a union of scholars from the natural and social sciences and the arts and humanities, and working with decision makers and the general public. The Millennium Alliance for Humanity and the Biosphere (MAHB),[81] which we already mentioned and to which both of us belong, is striving to develop a new type of intelligence, *foresight intelligence*: the ability to look ahead and then implement behavioral, institutional, and cultural changes that appear necessary to ensure a sustainable and equitable future for all. It seeks to create a platform where this intelligence can be generated through the coordination of dedicated institutions and individuals in civil society. Whether it will be successful is an open question, but we are convinced that a near-total reorganization and rescaling of society is required if the human predicament is to be resolved. There are hopeful signs that citizen-generated social-engineering groups in Australia—like GetUp![82]—could also assist, although to date no dedicated organization for targeted foresight intelligence yet exists in Australia. There are other hints of hope. For example, many subnational political entities, primarily in cities, are starting to deal seriously with the big environmental problems like climate disruption. At a much larger scale, the "nature-friendly" state of California, under the leadership of governors Arnold Schwarzenegger and Jerry Brown, has worked hard to deal with climate change. California has relatively strict environmental regulations on industry; it is the ninth largest economy in the world and corporations want access to its markets, so they tend to accept them. Public attitudes and smart, responsible politicians can make a big difference. If the world had California's environmental ethic, it would not solve its perfect storm of problems, but it sure would be a hell of a lot closer to doing so.

The multiplicity of groups working on solutions is heartening, but one should never forget that they are uncoordinated and that strong forces oppose them, such as the Heartland and World Growth Institutes mentioned previously. To some, it might appear that the pluto-theocratic agents of greed are winning, but with each of the steps they take our planet back, we must endeavor to take two forward. For example, one such step would be a complete revision of the US tax system. The plutocrats like to lie through their Murdochian henchmen at *Fox News*, the editorial pages of the *Wall Street Journal*, and the like that 47% of Americans pay no income taxes, implying the rich carry the burden of supporting the government's programs. That figure relates only to *income* tax, and the implication that the income tax represents the total tax burden is a clever lie. The "payroll tax," which the king of plutocrats Ronald Reagan doubled on the basis of a lie, amounts to another income tax that hits the middle class and the poor disproportionately, as do an army of other regressive taxes, such as sales taxes, driver's license fees, and the like.[83] These lies are propagated with other whoppers long exposed by experience—such as the lies that raising taxes on the super-rich would hurt the economy, that lowering them would provide benefits that would trickle down to the poor, that reducing the size of government would create jobs, and that Medicare and Medicaid are the source of US economic woes.[84] Australia's situation demonstrates that these assertions are incorrect.

But even Australia's decidedly more socialist tax arrangements hide cryptic benefits for the wealthy. There are generous incentives for first home buyers, payments for having children, and the like. Up until recently, this "baby bonus" was provided without regard for the personal wealth of the parents. In the United States, the income tax system also encourages rather than discourages over-reproduction through per-child tax reductions. Australia also has incentives for additional private health care* and tax breaks for the aged, regardless of personal wealth. There are some encouraging moves forward to remove tax havens for the rich, such as a reduction in the amount one can sequester away from tax in superannuation schemes (retirement funding), but there is still a long way to go to make the Australian tax system more efficient and fairer.

The entire systems of representative government in both the United States and Australia (to say nothing of many other countries) also badly need reexamination and reorganization. The original idea of

*Private health care incentives have recently become means tested (scaled to salary).

the founding fathers of the United States would be that prosperous, white, male, often slave-holding[85] land owners would vote to send one of their group to Washington to represent their interests and vote on their behalf. The "group" in this case (for representatives) was thought to be a few tens of thousands of voters. It seemed to work pretty well at first (except, of course, if you were unlucky enough to be poor, a slave, or a woman), but recently two factors have made it much less effective. One of them the founding fathers recognized: population growth. The other they did not—the incredible technological revolution in communications.[86] A combination of the spreading of the franchise and (mostly) an amazing population explosion has made congressional representatives and senators ever less representative. Today the number of people "represented" by each congressperson is over 700,000—more than twentyfold what the founding fathers originally envisioned. Instead of each being a special person who, having the benefits of education and special knowledge available in the capital, will make wise decisions, the representatives tend to be rich people or their hirelings with no special knowledge. They mostly use their riches to employ modern communications to establish positions that will be popular with their constituents and those who funded them in the first place. It is a partial return to a sort of Athenian direct democracy, a direct rule by the rich. A big difference, however, is that when the American government decides to go to war, cowardly "leaders" like George W. Bush and Dick Cheney hide at home and let the innocent and mostly poor young be slaughtered. In ancient Greece, leaders like Pericles strapped on their swords and shields and fought. That would be a sight—soft and plump politicians slogging through the desert sands of Iraq or the mountains of Afghanistan with rifles, engaging the enemy.

Today much public focus is on the world's financial troubles. They are genuine in one sense, relatively trivial in another. Any "debt crisis"—like the contrived budget "emergency" of the Abbott government[87]—could be solved by agreements among people simply, say, deciding that no one could have an income of more than $250,000 per year and a net worth of more than $10 million (pick any numbers, the principle is the same). It would cause a lot of trouble, but in theory it could be done. If, however, everyone agreed that the carrying capacity of Earth should now be made 5 billion people leading an average US lifestyle, or that the climate should be stabilized as it is today, or that no one should be exposed to hormone-mimicking chemicals, the tasks would be impossible in both theory and practice.

We know that some proposed changes would work, just by comparing the governance actions in Australia and the United States. Australia does not waste precious time and energy on debating the merits of gun control. It just controls guns, and the gun-related death statistics prove that it works (at least, relative to America). Australia has public health care and does not question its merits—the country's life expectancy (eighty-five for females, eighty for males) speaks volumes for its success. Even though the Abbott government is attempting to erode these established and effective social programs, Australia is still miles ahead of the United States in that regard. While Australia also had a carbon price, it did demonstrate real reductions in emissions without—contrary to claims of the opposite—substantially increasing the average citizen's cost of living.[88] To speak of climate disruption, let alone establish a Department of Climate Change as did Australia, and actually implement carbon pricing of some sort would currently be political suicide in the United States. Australia, however, does spend far too much energy engaged in less urgent issues like gay marriage and public gambling laws, and risks being dragged into the political quagmire and plutocracy of the United States. If Australians want to avoid the mess in which the United States has mired itself, they will heed that warning. If the United States wants to improve its lot, looking to some Australian successes would not hurt.

As we mentioned earlier, it is clear to us that the capitalist system at the very least needs drastic redesigning. The present system, as sociologist William Robinson put it, "must continually expand or collapse" and is "fast reaching the ecological limits of its reproduction."[89] The economists' emphasis on growth needs to end and its emphasis on efficiency needs to be tempered with more focus on equitable distribution. Governments and cultures should be reshaped so that one of their most openly recognized, major missions is to regulate the marketplace so that most externalities are internalized for the good of society, and that excess production of children or goods are both key externalities, imposing costs on all of society that are not captured in market prices. Society must recognize openly that it is impossible to internalize all externalities, and that the phobia against planning must be terminated (markets cannot plan). Everyone should face the fact that old-time (neoclassical) capitalism and present-day transnational capitalism, like socialism and communism, simply have not and cannot generate the sustainable redistribution as well as the material and population shrinkage that are essential to create an environmentally sound and equitable global society. Of course, no governance system is perfect, but elements from each will be required if we are to survive even semi-intact as a society. The challenge is immense and unprecedented, and the constraints of time, horrifying. The dilemma is exacerbated by the plutocrats buying politicians and creating and paying for a powerful and effective disinformation machine programmed to lie about environmental threats. Overcoming that machine will require much cooperation, which will likely not be achieved without new institutions, especially those dealing with education and communication, and a broad increase in social justice. There are many critical questions to be considered and the answers to only a few are at hand, despite the self-satisfaction of the Earth's destroyers when they meet annually at Davos, Switzerland, to brag and enjoy their spoils at the World Economic Forum: a confederacy of the clueless.

Improve Trans(inter)national Governance Frameworks

We have focused heavily on the internal governance structures that are the cause of many environmental woes in Australia and the United States, but improvements in transnational governance also have to play a strong and increasing role in alleviating the planet's problems. This area is made especially difficult to analyze as the world is now well into a transition away from a nation-state system to a transnational one—but a transition that is subject to much controversy among those attempting to analyze it.[90] We mentioned in previous chapters that international agreements—covering everything from ozone and wetlands protection, greenhouse gas emissions targets, trade in endangered species to how many whales can be killed—are signed with apparent enthusiasm by both countries. Some of these have been almost incredibly successful. For example, there has been a worldwide reduction in chlorofluorohydrocarbons causing damage to the ozone layer, agreements for nuclear weapons non-proliferation, international treaties on wetlands protection, and, at least for a while, a successful stoppage of the ivory trade. However, others have failed, such as global greenhouse gas mitigation policies. The long and short of our argument here is that international agreements, accords, and targets only appear to function when the self-interests of the signatory countries and the transnational capitalist class that controls much of the world economy generally concur. When they do not, agreements fail. In other words, we require far more binding and consequential agreements run by international bodies with real teeth. If there are no incentives and, importantly, no penalties for non-compliance, then we stumble forward at our own assured peril. Self-interested brinkmanship that remains uncurtailed by strong international legislation will not prevent the coming storm. In many ways, the international situation today resembles a large-scale version of the issues confronting the thirteen North American colonies that led to the Federalist/anti-Federalist debates over the proposed US Constitution. In essence, the issue was how to retain valued distinctions among the units while allowing them to solve the problems requiring collective action and while retaining the dominance of propertied white men—the trans-colonial capitalist elite. Examples then were how to provide for the common defense if the British returned and how generally to control foreign affairs and trade. Today analogous issues are how to deal with global environmental problems and suppress nuclear war.

Revamping Education

Cultural evolution will obviously need to be accelerated in the right direction before any kind of real progress can be made. That would seem to require changes in the education system, moving the culture more toward a basis in evidence rather than in faith. Indeed, it has been shown repeatedly that increasing general levels of education increases participation in environmental movements,[91] and the type of education is essential in this process.[92] But changing education meaningfully would be no trivial task—it would involve speeding a trend already in place, but moving unevenly and usually at a snail's pace. If, however, the urgency of the situation could be clarified for more of the population, it seems possible that progress could be speeded substantially. Ideally the major anomalies, like the co-location of universities and chapels, would start to disappear.

We do not think that large-scale redistribution of wealth and environmental responsibility will be easy in any case, but a thorough revision of the public education system from preschool up[93] would go a long way toward providing solutions to overconsumption and encouraging the needed change of norms. If any nations have the ability to make such changes, it is the prosperous and literate ones like Australia and the United States. It would also enhance the status of women and the poor—reducing inherited advantage of all kinds. In the long term, it should make solving other environmental problems much easier. The dismal state of macroeconomics today demonstrates the importance of making how humanity interacts with its environment an integral part of education.

The need to drag universities into the twenty-first century is manifest in their departmental structure, some of which traces back to Aristotle[94] and to the Royal Society in 1664. Here we return to the key "discipline" of economics, since it deals with the allocation of scarce resources and perhaps is the most influential in policy circles but is, regrettably, largely isolated from reality. Much of the human predicament traces to the organization of the global economy, the resultant power relationships, and a black box called "technological change" in economic growth models that can apparently allow some humans to ignore the laws of nature.

A good place to start reorganizing universities would be to reconsider how the crucial topic of economics is taught. David Lam wrote:

> I'm optimistic not because the problems posed by continued population growth are simple or because they will take care of themselves, but rather because the last 50 years have demonstrated our capacity to recognize the challenges and to tackle them with hard work and creativity.[95]

Ironically, part of today's dilemma is the economic system's (and Lam's) failure to recognize the environmental challenges. If all economists were exposed early on to ecological analysis[96] or read books like this one, the over-optimism just expressed would disappear, or at least wane. Human efforts have generated a population size and degree of consumption already well beyond what the planet can support indefinitely, even with at least 2 billion people condemned to utter poverty. If faculty members could be reassigned to new units within restructured universities, students could learn to apply economics to fundamental human problems, a partial return to the political economy of the eighteenth century. For example, perhaps resource economists could be associated with fisheries biologists, and behavioral economists with political scientists or philosophers. Then we all might no longer have to suffer the pronouncements of *Wall Street Journal* economists, politicians, and pundits treating economic growth among the rich of 3–5% annually as a necessary norm rather than the gargantuan threat to the persistence of civilization it actually is. Let them eat cake, indeed.

Fixing the education system is necessary, but the education of adults is needed to transform economic thinking rapidly by adopting "green accounting," in which the flawed indicator of GDP is replaced by others that pay attention to the state of natural, human, and social capital.[97] There are encouraging efforts in this direction, such as the already-mentioned Natural Capital Project,[98] but they must be rapidly amplified. Obviously, it is not just norms related to consumption that need changing—so do those related to the rights of women and thus reproduction. Evolving norms will be key, and with appropriate techniques, this can be done carefully and humanely.

Consider, for example, efforts to employ an approach rooted in social psychology: the "*Sabido* Method." It has been used successfully to increase the acceptability of family planning through serialized dramas on television and radio. The programs have captivated audiences while imparting socially beneficial values. They provide the audience with a range of characters—pseudo-kin (non-relatives perceived as relatives)—to whom people relate and who are involved in various dilemmas. Through the interactions of the actors and the twists and turns in the plot, realistic solutions were explored. At the end of the drama, the main character decides to have only two children. *Sabido* programming related to women's rights and family planning has been shown throughout Africa, Asia, and Latin America. After one serial program, *Televisa*, was broadcast in Mexico, that government's national population council (CONAPO) reported large increases in the numbers of people seeking family-planning information and of women volunteering to work in the national program of family planning. Contraceptive sales increased by almost a quarter in one year, and more than half a million women enrolled in family-planning clinics, an increase of 33% (compared to a 1% decrease the previous year). Several soap operas with family-planning themes were on the air in Mexico between 1977 and 1986, and its population growth rate dropped by a third. Surveys showed Mexicans attributed the changed norms to the soap operas.

Separate Religion and State

As we have pointed out, religion has no place in rational governance. Making firm commitments to separate religious (faith-based) beliefs from evidence-based policies would be a major leap forward for the US political system and possibly save Australia the embarrassment and heartache of theocratic political regression. The US federal constitutional separation of church and state has been reasonably successful in keeping religious indoctrination out of public schools, although not out of politics, where elected officials often resort to pleading with imaginary beings. Australia, in contrast, has state control of religion in schools and the "divinely led" government of Tony "the Monk" Abbott appears hell-bent on promulgating fear of the supernatural therein. Australia could take at least a constitutional page out of the United States' book by making church and state officially decoupled, while simultaneously avoiding the moronic US tradition of using faith to set government policy. Likewise, our training centers (universities and the like) are no place for the oxymoron that is *theology*—removing such structural barriers to intelligent governance would be part of the educational overhaul. Denis Diderot, the French Enlightenment philosopher and encyclopedist, had this figured out more than two centuries ago:

> Wandering in a vast forest at night, I have only a faint light to guide me. A stranger appears and says to me: "My friend, you should blow out your candle in order to find your way more clearly." This stranger is a theologian.[99]

In view of the great recent successes of those promoting the war on science, it is clear that the Endarkenment is upon us. To counter this, scientists need to strive to become modern Voltaires and take a leading role in mobilizing and informing the public. Otherwise humanity will likely continue on its business-as-usual, faith-based trajectory to collapse. To see the growing results of that lethal course, turn on your television news or travel anywhere. *Lector, si monumentum requiris, circumspice.**

Laud the Good Guys

We have gone out of our way to point a finger at some of the worst of the bad guys†—the plutocrats who believe corporations are individuals, the pundits who hide their greed agendas behind fables of societal improvement, the theocrats who manipulate and lobotomize their sheep with promises of divine intervention, the "intellectuals" whose platforms are based on inconsistencies, bad data, or outright lies—but we have not yet mentioned many of the "good guys." In a world increasingly hijacked by the greedy, the good guys are relatively difficult to identify, but they do exist. Some are principled, rational people swimming hard against the current of lunacy, and so their efforts might go unnoticed or underappreciated. Among these we want to mention, people like Al Gore, who, while much maligned, began changing the face of climate-disruption politics the world over. Tim Wirth, former US senator from Colorado, was a pillar of commitment to environmental issues, and Jerry Brown and Arnold Schwarzenegger are a few modern-day politicians who put at least one US state on the map in terms of progressive environmental policy. In Australia Bob Carr—the former premier of New South Wales and former foreign minister for the Gillard Labor government—has been a bastion of environmental protection in Australian politics for nearly twenty years. The list is not limited to politicians—even some of the corporate world's biggest successes are committed to improving this planet. From Melinda and Bill Gates and their work on improving the health of the poor, to Australian electronics magnate Dick Smith, who speaks boldly on the issue of Australian overpopulation, there are extremely wealthy people—some of whom we know well—who have put substantial amounts of their good fortune and their time into making society a better place.

*"Reader, if you seek a monument, look around you."

†Gals, too, of course.

End the History Hangover

Humanity today must simultaneously learn from the past and yet forgive historical grievances. Yes, P.R.E. lives on land once occupied by Ohlone Native Americans, who were largely exterminated by European impacts, and C.J.A.B. owns property on lands once occupied by the Peramangk people of the Adelaide Hills region of South Australia. In fact, almost everyone can look to the past and claim that his or her group got a raw deal (and suffered some "ethnic cleansing") at one time or another. Jews suffered horrifically at the hands of Christians, Christians in the first few centuries AD were frequently thumped by the Romans, and, subsequently, various subcults took up slaughtering each other in great numbers. Palestinians have not been treated well by Christians (remember the Crusades) and continue to get pummeled by the Israelis. Then there was the rape of Nanking by the Japanese army in 1937, the US atom bombs against the Japanese in 1945, the defeat of the Serbians at the first Battle of Kosovo in 1389 and the "ethnic cleansing" of 1998–99, the deaths of millions on both sides during the partition of India in 1947, the murder of thousands in the 2001 World Trade Center in reprisal for the US presence in Saudi Arabia, the slaughter of "heretic" Christians by other Christians in the thirteenth-century Albigensian Crusade, and so on, *ad nauseam*.

People have not treated those they viewed as "others" well, going back as far as we have historical or archaeological evidence. The excuses given for continuing conflict often border on the ludicrous. How often are we treated to scenes like that of an Israeli woman being forcibly removed from an illegal settlement, screaming that the land had been given to Jews millennia earlier by their favorite supernatural sponsor? How about the wonderful folks from Kansas's Westboro Baptist Church parading with signs that say "God hates fags." The time has come for humanity as a whole to lessen its focus on family issues from the past and look forward to how it can unite as a single family at a higher level to work cooperatively for a peaceful, equitable, and sustainable civilization. We cannot change the awful past, but we could avoid an incalculably awful future.

Australia and the United States have some potential in this area because as relatively "young" nations, their historical baggage should be lighter than in Old World countries. The slaughter, maltreatment, and suppression of their indigenous populations notwithstanding, looking forward without the crutch of past aggressions and grudges could give both countries a decided advantage in taking leading roles in environmental planning and reconciliation. Yes, the Palestinians got a raw deal when Israel was created for Jews, who got an even rawer deal in Europe (and were prevented from being absorbed into the United States by American anti-Semitism). But it is time both the Jews and the Arabs got over it and worked together to build a community that finds ways of persisting together in the overpopulated and water-short land of Palestine. If chimps can reconcile after disputes, surely people should be better at it.

The United States could go all out for non-polluting energy, dissolve the US military's Central and African Commands and all its other military establishments designed to control resource flows (most importantly, petroleum), and greatly increase aid to both groups. Australia could pitch in and perhaps make a major effort to help both China and India to transition to more solar power, first of all by mandating escalation of coal prices to put coal mining out of business in Australia in five years. These pie-in-the-sky suggestions show the sorts of things that *could* be done—but we are not holding our breath!

Some Parting Words

After casting a wide net of recommendations, the question of whether we think much can be changed should be asked. The short answer is "no." Greed is not about to disappear—even the bad economists know that; if there is a way to exploit a system for personal gain, there will always be a willing punter.* Nor is religious fervor about to wane in some unlikely second Renaissance of reason. If anything, we expect religious fundamentalism to grow. There will always be theocrats posing as "people for the people," and plutocrats trying to skim more cream off the top of society's milk bottle than they should. We are talking, after all, about human beings.

Likewise, an environmental revolution is not just around the corner, and a sustainability utopia will not emerge as a consequence of our pleas. The simple fact that there are already far, far too many people

*A "customer" ready to try.

on Earth, virtually all of whom are looking to advance their personal situation, means that in the absence of suddenly beaming the majority of us up to some undiscovered planet, ecosystems will continue to degrade over the coming centuries. If only Douglas Adams's fantasies about transporting some humans to another planet[100] were even partially true! Throw in the amount of climate disruption to which the world is already committed, and the future will feature some particularly nasty outcomes for the environment, humanity, and everything.

So in the tradition of striving for scientific objectivity, we remain pessimists. But as we are often asked at public-speaking events, why even continue to do what we do? Why do we even bother? The simplest answer, the way we see things is that if we do not, our world will come crashing down around us much faster than it would otherwise. If we can give our offspring and their families and our friends just another decade without resource wars, without burning horribly in a bushfire, or without dying from some nasty disease, we will damn well try. If we keep challenging people to do something about it, maybe some will eventually rise to the challenge. So while we have plenty to say about what is wrong with the world, and in particular in the United States and Australia, all of our recommendations boil down to this: Do not let the bastards get away with it!

What we mean is that it is the duty of every rational citizen to call out the insanities, the ridiculous statements, the self-serving political actions, and the stupid justifications provided for screwing over our only home. Too often we sit at home and shake our heads while some complete idiot spouts inane words on the televised news program or in the newspaper. We complain to our colleagues in the office tearoom about what so-and-so politician did in the latest bill-voting session. We groan, and even grind our teeth in anger, when some theocrat tries to tell educated adults that a woman does not have the right to govern her own vagina. But that is usually where we leave it. While we tend to exercise our democratic right to choose our representatives by voting, there is so, so much more we can do to influence policy.

Scientists are among the worst offenders at dithering in this arena. Many, perhaps most, of our colleagues hide behind the veil of "scientific impartiality" that they somehow think will be compromised if they so much as show an opinion.[101] Yet scientists are supposed to be the most informed of us all! Why are there not weekly demonstrations of Australian and American scientists marching along the streets of Canberra and Washington demanding that the lunatics occupying our respective houses of government terminate their inane positions? Why are there not more of us singing from the proverbial rooftops that our environmental predicament is so much, much worse than most people realize? Why don't major ecological and conservation societies take strong stands on the obvious drivers of environmental destruction such as overpopulation, overconsumption, and use of fossil fuels? Only in Canada so far has the government's anti-scientific efforts led scientists to take to the streets.[102]

Yes, there are exceptions to those scientists who sit back and do little or nothing. To name a few of our friends: Jim Brown, Partha Dasgupta, Carl Folke, John Holdren, Bill Laurance, Tom Lovejoy, Norman Myers, Daniel Pauly, Stuart Pimm, Peter Raven, Harry Recher, Michael Soulé, Frank Talbot, Ed Wilson, George Woodwell, and many others have been warning about overpopulation and the extermination of biodiversity for decades. Tony Barnosky, Andy Beattie, Barry Brook, Gerardo Ceballos, Brian Czech, Gretchen Daily, Jared Diamond, Rodolfo Dirzo, Chris Field, Tim Flannery, Liz Hadly, Charlie Hall, Clive Hamilton, John and Mel Harte, Lesley Hughes, Simon Levin, Jack Liu, Jane Lubchenco, Mike Mann, Hal Mooney, Michael Oppenheimer, Camille Parmesan, Terry Root, Ben Santer, the late Steve Schneider, Tom Wigley, and many others have struggled to alert humanity to the perils of climate disruption and other elements of the perfect storm. James Hansen, the famous NASA climatologist, has gotten himself arrested for demonstrating against backward energy policies. Pete Myers, in the footsteps of Rachel Carson, has been at the front lines of trying to inform people of the dangers of toxification. So has Tyrone Hayes, at great personal risk. Other colleagues demonstrate against the transport of coal,[103] the establishment of a new mine, or the proposal to cut down a forest. But these few cries are drowned out by the silence of the majority.

We are certainly not calling for all scientists to abandon their day jobs and become full-time activists, but we are demanding that they call a spade a spade when one rears its ugly head in public policy. What of the rest of you? Most of you reading this book are probably not scientists, but you obviously have engaged brains and are interested, even if only peripherally, in the problems we have discussed in this book. Otherwise, you would not be reading these final words. The first step is, of course, to inform yourself of the facts and not swallow the drivel spewed from the pundits' various orifices. Hopefully this book has given you at least a taste for that knowledge and the urge to check up on us and see if we have a hidden agenda. The next step is to do exactly what we prescribe to scientists. Get out there and let the society-killing miscreants know that you will not stand for it. Only then will we have any chance of giving our grandchildren a happy, healthy, just, and fulfilling life.

Acknowledgments

We are grateful for the feedback received on the text from Karole Armitage, Andrew Beattie, Barry Brook, Sarah Carlson, Lisa Daniel, Tim Daniel, Joan Diamond, Anne Ehrlich, Larry Goulder, Sandra Kahn, William Laurance, Thomas Lovejoy, Chase Mendenhall, Simon Nasht, Kirk Smith, Susan Thomas, Chris Turnbull, Peter Ward, Karah Wertz, Paul Willis, and Adam Wynn. We also thank other colleagues for contributing previously unpublished content, including William Bond, Jon Foley, Larissa Jarvis, Tony Peacock, Dany Plouffe, Navin Ramankutty, and Euan Ritchie.

Corey thanks the University of Adelaide for financial support; Stanford University for a place to think and write; Janet Elder for technical support; Lana and Gene for their hospitality, support, and encouragement; K for giving him the opportunities to do such extracurricular things as write books when there are clearly more important things to do; little C for her unrelenting affection and smiles, despite having no idea why all this was necessary.

Paul thanks Stanford University for putting up with him as an active faculty member for fifty-four years, and Janet Elder for the same for three years. Anne Ehrlich has been extremely helpful in working on the manuscript and has repeatedly told me it was probably too frank. Maybe so, but I believe we owe it to our great-grandchildren and the wonderful kids of friends (like little C) to tell it like it is. I am also deeply indebted to Pete and Helen Bing, Larry Condon, and Wren and Tim Wirth for their friendship and support. The book is written in the spirit of my late friends: Loy Bilderback, Stan and Marion Herzstein, Dick Holm, LuEsther Mertz, John Montgomery, Steve Schneider, and John Thomas.

You deserve to know a little about our backgrounds. C.J.A.B. was born in Canada, but after many years Down Under is now a "true blue"

Aussie, having been tempered by sweating years in the climatic hell of Darwin and nearly going "troppo,"* shivering for months in the winter drizzle of Hobart, and finally ending up in the Mediterranean bliss† of Adelaide. P.R.E. was born in the awful East of the United States and served five years in Kansas before fetching up in California, the land of milk and honey. He has spent roughly three years as a Sydney-sider and has visited almost all of Australia with the exception of a few outback areas. Indeed, he has seen much more of Australia than most Australians.

*Crazy due to the intense, unrelenting (tropical) heat.

†Which of course means, *inter alia*, plenty of delicious wine.

Notes

PREFACE

1. From the *Australia Women's Weekly*, February 12, 1941, trove.nla.gov.au/ndp/del/article/47208821 (accessed September 30, 2012).

2. Australian Associated Press, "Julie Bishop Backs US as Key Ally and Superpower Despite China's Rise," June 18, 2014, www.theguardian.com/world/2014/jun/18/julie-bishop-backs-us-as-key-ally-and-superpower-despite-chinas-rise (accessed June 19, 2014).

3. Ngeringa Vineyards, www.ngeringa.com (accessed August 4, 2014).

PROLOGUE

1. N. Klein, *The Shock Doctrine: The Rise of Disaster Capitalism* (New York: Picador, 2007).

CHAPTER ONE

1. P. R. Ehrlich, *The Population Bomb* (New York: Ballantine Books, 1968).

2. CIA, *The World Factbook*, www.cia.gov/library/publications/the-world-factbook/ (accessed September 18, 2014).

3. J. Salatin, *Folks, This Ain't Normal* (New York: Center Street, 2011).

4. Australian Bureau of Statistics, www.abs.gov.au (accessed October 20, 2012).

5. S. Rubinsztein-Dunlop, "Australia's Prison System Overcrowded to Bursting Point with More than 33,000 People in Jail," July 2, 2014, www.abc.net.au/news/2014-07-02/austrlaian-prison-overcrowding-female-populations-growing/5567610 (accessed July 2, 2014).

6. "A Stinger for Antonin." *Economist*, July 30, 2012, www.economist.com/blogs/democracyinamerica/2012/07/gun-rights (accessed July 15, 2014).

7. J. Turley, "Rocket Launchers and the Second Amendment," July 29, 2012, jonathanturley.org/2012/07/29/rocket-launchers-and-the-second-amendment (accessed June 19, 2014).

8. "Medicine: the Thalidomide disaster." *Time Magazine*, August 10, 1962, www.time.com/time/magazine/article/0,9171,873697-1,00.html (accessed August 20, 2012).

9. E. Alterman, *The Nation* 10 (2011).

10. N. Klein, *The Shock Doctrine: The Rise of Disaster Capitalism* (New York: Picador, 2007).

CHAPTER TWO

1. S. M. Coupe, *Historic Australia* (Sydney: Golden Press, 1982).

2. D. S. Trigger, *Whitefella Comin': Aboriginal Responses to Colonialism in Northern Australia* (Cambridge: Cambridge University Press, 1992).

3. E. Callaway, "Studies Slow the Human DNA Clock," *Nature* 489 (2012): 343–44.

4. J. M. Diamond, *Guns, Germs, and Steel: The Fates of Human Societies* (New York: Norton, 1997).

5. J. T. Clark, "Early Settlement in the Indo-Pacific," *Journal of Anthropological Archaeology* 10 (1991): 27–53.

6. Diamond, *Guns, Germs, and Steel.*

7. C. J. A. Bradshaw, "Little Left to Lose: Deforestation and Forest Degradation in Australia since European Colonization," *Journal of Plant Ecology* 5 (2012): 109–20.

8. J. C. Crowell and L. A. Frakes, "Late Palaeozoic Glaciation of Australia," *Journal of the Geological Society of Australia* 17 (1971): 115–55.

9. M. R. Raupach, J. M. Kirby, D. J. Barrett, and P. R. Briggs, *Balances of Water, Carbon, Nitrogen and Phosphorus in Australian Landscapes: (1) Project Description and Results* (Canberra: CSIRO Land and Water, 2001), 1–37.

10. S. Lane, *Aboriginal Stone Structures in Southwestern Victoria* (Melbourne: Aboriginal Affairs Victoria, 2009).

11. B. Pascoe, *Dark Emu, Black Seeds: Agriculture or Accident?* (Broome, Western Australia: Magabala Books, 2014).

12. H. K. Gibbs et al., "Tropical Forests Were the Primary Sources of New Agricultural Land in the 1980s and 1990s," *Proceedings of the National Academy of Sciences of the USA* 107 (2010): 16732–37.

13. C. J. A. Bradshaw, N. S. Sodhi, and B. W. Brook, "Tropical Turmoil—a Biodiversity Tragedy in Progress," *Frontiers in Ecology and the Environment* 7 (2009): 79–87.

14. C. N. Johnson, *Australia's Mammal Extinctions: A 50,000 Year History* (Cambridge: Cambridge University Press, 2006).

15. A. J. Pitman, G. T. Narisma, R. A. Pielke, and N. J. Holbrook, "Impact of Land Cover Change on the Climate of Southwest Western Australia," *Journal of Geophysical Research: Atmospheres* 109 (2004): D18.

16. B. W. Brook and C. N. Johnson, "Selective Hunting of Juveniles as a Cause of the Imperceptible Overkill of the Australian Pleistocene Megafauna," *Alcheringa: An Australasian Journal of Palaeontology* 30 (2006): 39–48.

17. R. Bliege Bird, D. W. Bird, B. F. Codding, C. H. Parker, and J. H. Jones, "The 'Fire Stick Farming' Hypothesis: Australian Aboriginal Foraging Strategies, Biodiversity, and Anthropogenic Fire Mosaics," *Proceedings of the National Academy of Sciences of the USA* 105 (2008): 14796–801.

18. G. Miller et al., "Sensitivity of the Australian Monsoon to Insolation and Vegetation: Implications for Human Impact on Continental Moisture Balance," *Geology* 33 (2005): 65–68.

19. R. A. Bradstock, A. M. Gill, and R. J. Williams, eds., *Flammable Australia: Fire Regimes, Biodiversity and Ecosystems in a Changing World* (Collingwood, Victoria, Australia: CSIRO Publishing, 2012).

20. J. C. Z. Woinarski et al., "The Disappearing Mammal Fauna of Northern Australia: Context, Cause, and Response," *Conservation Letters* 4 (2011): 192–201.

21. T. Flannery, *The Eternal Frontier: An Ecological History of North America and Its Peoples* (London: William Heinemann, 2001).

22. B. W. Brook, N. S. Sodhi, and C. J. A. Bradshaw, "Synergies among Extinction Drivers Under Global Change," *Trends in Ecology and Evolution* 25 (2008): 453–60.

23. Flannery, *The Eternal Frontier.*

24. Ibid.

25. J. Diamond, *Collapse: How Societies Choose to Fail or Succeed* (New York: Viking Press, 2005).

26. Ibid.

27. Diamond, *Guns, Germs, and Steel*; Diamond, *Collapse.*

28. G. Grandin, *Empire of Necessity: Slavery, Freedom, and Deception in the New World* (New York: Metropolitan Books, 2014).

29. M. Kurlansky, *Cod: A Biography of the Fish That Changed the World* (New York: Walker and Co., 1997).

30. World Bank, "Data: Rail lines," data.worldbank.org/indicator/IS.RRS.TOTL.KM (accessed June 25, 2014).

31. Ibid.

32. United States Census Bureau, www.census.gov (accessed June 25, 2014).

33. C. J. A. Bradshaw et al., "Brave New Green World—Consequences of a Carbon Economy for the Conservation of Australian Biodiversity," *Biological Conservation* 161 (2013): 71–90.

34. John Vidal, "The State of Crop Harvests Around the World," *Guardian*, October 10, 2012, www.guardian.co.uk/environment/2012/oct/10/crop-harvests-world (accessed February 11, 2015).

35. Bradshaw, "Little Left to Lose."

36. Australian Bureau of Statistics, www.abs.gov.au/ausstats/abs@.nsf/mf/3101.0 (accessed September 15, 2014).

CHAPTER THREE

1. A. Leopold, *A Sand County Almanac* (New York: Ballantine, 1949).

2. Michigan State University, Animal Legal & Historical Center, www.animallaw.info/statutes/stusfd16usca3371.htm (accessed August 20, 2012).

3. K. C. Armitage, *The Nature Study Movement: The Forgotten Popularizer of America's Conservation Ethic* (Lawrence: University Press of Kansas, 2009).

4. E. Seton, T. *Lives of the Hunted* (New York: Charles Scribner's Sons, 1901).

5. G. Stratton-Porter, *Moths of the Limberlost* (Garden City, NY: Doubleday, Page & Company, 1916).

6. A. Kallet and F. J. Schlink, *100,000,000 Guinea Pigs: Dangers in Everyday Foods, Drugs, and Cosmetics* (New York: Vanguard Press, 1933).

7. M. Vasilogambros and S. Mimms, "Scalia Defends Citizens United Decision," *National Journal*, July 18, 2012, www.nationaljournal.com/politics/scalia-defends-citizens-united-decision-reflects-on-term-in-rare-tv-appearance-20120718 (accessed August 20, 2012).

8. D. Milbank and J. Blum, "Document Says Oil Chiefs Met with Cheney Task Force," *Washington Post*, November 16, 2005, www.washingtonpost.com/wp-dyn/content/article/2005/11/15/AR2005111501842.html (accessed August 20, 2012).

9. R. Carson, *Silent Spring* (Boston: Houghton Mifflin, 1962).

10. "Medicine: The Thalidomide Disaster," *Time Magazine*, August 10, 1962, www.time.com/time/magazine/article/0,9171,873697-1,00.html (accessed August 20, 2012).

11. G. Speth, "The Global 2000 Report to the President," *Boston College Environmental Affairs Law Review* 8 (1980): 695–703.

12. "A Look Back at Reagan's Environmental Record," *Grist*, June 11, 2004, grist.org/article/griscom-reagan (accessed August 20, 2012).

13. J. Koshin, "Shifting Visions: Developmentalism and Environmentalism in Australian History," *Australian Studies* 3, article 2226 (2011).

14. Ibid.

15. L. W. Braithwaite, "Conservation of Arboreal Herbivores: The Australian Scene," *Australian Journal of Ecology* 21 (1996): 21–30.

16. C. J. A. Bradshaw, "Little Left to Lose: Deforestation and Forest Degradation in Australia since European Colonization," *Journal of Plant Ecology* 5 (2012): 109–20.

17. Koshin, "Shifting Visions."

18. T. Griffiths, "How Many Trees Make a Forest?: Cultural Debates about Vegetation Change in Australia," *Australian Journal of Botany* 50 (2002): 375–89.

19. S. Feary, in *Australia and New Zealand Forest Histories*, ed. J. Dargavel (Kingston, A.C.T., Australia: Australian Forest History Society Inc., 2005), 9–15.

20. J. Dargavel, in *Australia and New Zealand Forest Histories*, ed. J. Dargavel (Kingston, A.C.T., Australia: Australian Forest History Society Inc., 2005), 25–33.

21. C. Jordan, "Progress Versus the Picturesque: White Women and the Aesthetics of Environmentalism in Colonial Australia, 1820–1860," *Art History* 25 (2002): 341–57.

22. Griffiths, "How Many Trees Make a Forest?"; Dargavel, in *Australia and New Zealand Forest Histories*.

23. B. J. Stubbs, in *Australia and New Zealand Forest Histories*, ed. J. Dargavel (Kingston, A.C.T., Australia: Australian Forest History Society Inc., 2005), 34–41.

24. Ibid.

25. Koshin, "Shifting Visions."

26. C. Rootes, "Environmentalism in Australia," *Environmental Politics* 10 (2001): 134–39.

27. G. De'ath, K. E. Fabricius, H. Sweatman, and M. Puotinen, "The 27-Year Decline of Coral Cover on the Great Barrier Reef and Its Causes," *Proceedings of the National Academy of Sciences of the USA* 109 (2012): 17995–99.

28. Stubbs, in *Australia and New Zealand Forest Histories.*

29. Koshin, "Shifting Visions."

30. R. Routley and V. Routley, *The Fight for the Forests: The Takeover of Australian Forests for Pines, Wood Chips and Intensive Forestry* (Canberra: Australian National University, 1973).

31. K. J. Frawley, in *Australia and New Zealand Forest Histories,* ed. J. Dargavel) (Kingston, A.C.T., Australia: Australian Forest History Society Inc., 2005), 42–48.

32. Stubbs, in *Australia and New Zealand Forest Histories.*

33. J. E. M. Watson et al., "Wilderness and Future Conservation Priorities in Australia," *Diversity and Distributions* 15 (2009): 1028–36.

34. R. L. Pressey, "Conservation Reserves in NSW: Crown Jewels or Leftovers," *Search* 26 (1995): 47–51.

35. A. Conacher, "Review Essay: Changing Environmentalism," *Australian Geographer* 23 (1992): 177–83.

36. A. Davison, in *Second State of Australian Cities Conference 30 Nov–02 Dec 2005.*

37. Rootes, "Environmentalism in Australia."

38. A. S. K. Frank et al., "Experimental Evidence That Feral Cats Cause Local Extirpation of Small Mammals in Australia's Tropical Savannas," *Journal of Applied Ecology* 51 (2014): 1486–93.

CHAPTER FOUR

1. J. Maclaurin and K. Sterelny, *What Is Biodiversity?* (Chicago: University of Chicago Press, 2008).

2. P. J. Crutzen, "Geology of Mankind: The Anthropocene," *Nature* 415 (2002): 23.

3. D. M. Raup, "Biological Extinction in Earth History," *Science* 231 (1986): 1528–33; D. M. Raup, "The Role of Extinction in Evolution," *Proceedings of the National Academy of Sciences of the USA* 91 (1994): 6758–63.

4. M. J. Benton, *When Life Nearly Died: The Greatest Mass Extinction of All Time* (London: Thames & Hudson, 2003).

5. S. L. Pimm et al., "The Biodiversity of Species and Their Rates of Extinction, Distribution, and Protection," *Science* 344, doi:10.1126/science.1246752 (2014).

6. World Wildlife Fund, "Living Planet Report 2014," www.worldwildlife.org/pages/living-planet-report-2014 (accessed October 21, 2014).

7. M. J. Costello, R. M. May, and N. E. Stork, "Can We Name Earth's Species Before They Go Extinct?" *Science* 339 (2013): 413–16.

8. Ibid.; C. Mora, D. P. Tittensor, S. Adl, A. G. B. Simpson, and B. Worm, "How Many Species Are There on Earth and in the Ocean?" *PLoS Biology* 9, e1001127 (2011).

9. Costello, May, and Stork, "Can We Name Earth's Species Before They Go Extinct?"

10. P. R. Ehrlich, *World of Wounds: Ecologists and the Human Dilemma* (Oldendorf/Luhe, Germany: Ecology Institute, 1997).

11. D. L. Roberts and A. R. Solow, "Flightless Birds: When Did the Dodo Become Extinct?" *Nature* 426 (2003): 245.

12. B. W. Brook, N. S. Sodhi, and C. J. A. Bradshaw, "Synergies among extinction drivers under global change," *Trends in Ecology and Evolution* 25 (2008): 453–60.

13. W. F. Laurance et al., "A Global Strategy for Road Building," *Nature* 513 (2014): 229–32.

14. D. S. Trigger, *Whitefella Comin': Aboriginal Responses to Colonialism in Northern Australia* (Cambridge: Cambridge University Press, 1992).

15. C. J. A. Bradshaw, N. S. Sodhi, K. S. H. Peh, and B. W. Brook, "Global Evidence That Deforestation Amplifies Flood Risk and Severity in the Developing World," *Global Change Biology* 13 (2007): 2379–95.

16. Y. Pan et al., "A Large and Persistent Carbon Sink in the World's Forests," *Science* 333 (2011): 988–93.

17. S. E. Sexton, Z. Lei, and D. Zilberman, "The Economics of Pesticides and Pest Control," *International Review of Environmental and Resource Economics* 1 (2007): 271–326.

18. W. F. Laurance et al., "Improving the Performance of the Roundtable on Sustainable Palm Oil for Nature Conservation," *Conservation Biology* 24 (2010): 377–81.

19. I. C. Field, M. G. Meekan, R. C. Buckworth, C. J. A. Bradshaw, "Susceptibility of Sharks, Rays and Chimaeras to Global Extinction," *Advances in Marine Biology* 56 (2009): 275–363.

20. "Man Bites Shark," *ConservationBytes.com*, January 7, 2009, conservationbytes.com/2009/01/07/man-bites-shark/ (accessed September 15, 2014).

CHAPTER FIVE

1. J. D. Dingell, in *Balancing on the Brink of Extinction: The Endangered Species Act and Lessons for the Future*, ed. K. A. Kohm (Washington, DC: Island Press, 1991), 25–30.

2. C. J. A. Bradshaw, "Little Left to Lose: Deforestation and Forest Degradation in Australia since European Colonization," *Journal of Plant Ecology* 5 (2012): 109–20.

3. A. Singh, H. Shi, T. Foresman, and E. A. Fosnight, "Status of the World's Remaining Closed Forests: An Assessment Using Satellite Data and Policy Options," *Ambio* 30 (2001): 67–69.

4. Bradshaw, "Little Left to Lose."

5. Ibid.

6. N. Myers, R. A. Mittermeier, C. G. Mittermeier, G. A. B. da Fonseca, and J. Kent, "Biodiversity Hotspots for Conservation Priorities," *Nature* 403 (2000): 853–58.

7. Critical Ecosystem Partnership Fund, "The Biodiversity Hotspots," www.cepf.net/resources/hotspots (accessed March 16, 2015).

8. A. J. Pitman, G. T. Narisma, R. A. Pielke, and N. J. Holbrook, "Impact of Land Cover Change on the Climate of Southwest Western Australia," *Journal of Geophysi-*

cal Research: Atmospheres 109 (2004): D18; G. T. Narisma and A. J. Pitman, "The Impact of 200 Years of Land Cover Change on the Australian Near-Surface Climate," *Journal of Hydrometeorology* 4 (2003): 424–36; G. T. Narisma and A. J. Pitman, "Exploring the Sensitivity of the Australian Climate to Regional Land-Cover-Change Scenarios under Increasing CO_2 Concentrations and Warmer Temperatures," *Earth Interactions* 10 (2006): 1–27; R. C. Deo, "Links between Native Forest and Climate in Australia," *Weather* 66 (2011): 64–69; R. C. Deo et al., "Impact of Historical Land Cover Change on Daily Indices of Climate Extremes Including Droughts in Eastern Australia," *Geophysical Research Letters* 36 (2009).

9. E. Lepers et al., "A synthesis of information on rapid land-cover change for the period 1981–2000," *BioScience* 55 (2005): 115–24.

10. "South Australia's tattered environmental remains," *ConservationBytes.com*, April 16, 2014, conservationbytes.com/2014/04/16/south-australias-tattered-environmental-remains/ (accessed July 3, 2014).

11. D. Pimentel, R. Zuniga, and D. Morrison, "Update on the Environmental and Economic Costs Associated with Alien-Invasive Species in the United States," *Ecological Economics* 52 (2005): 273–88.

12. C. N. Johnson, *Australia's Mammal Extinctions: A 50,000 Year History* (Cambridge: Cambridge University Press, 2006).

13. A. S. K. Frank et al., "Experimental Evidence That Feral Cats Cause Local Extirpation of Small Mammals in Australia's Tropical Savannas," *Journal of Applied Ecology* 51 (2014): 1486–93.

14. W. F. Laurance et al., "Averting Biodiversity Collapse in Tropical Forest Protected Areas," *Nature* 489 (2012): 290–94.

15. Ibid.; J. Geldmann et al., "Effectiveness of Terrestrial Protected Areas in Reducing Habitat Loss and Population Declines," *Biological Conservation* 161 (2013): 230–38.

16. D. P. McCarthy et al., "Financial Costs of Meeting Global Biodiversity Conservation Targets: Current Spending and Unmet Needs," *Science* 338 (2012): 946–49.

17. A. Rourke, "Australian Coroner Finds That Dingo Took Baby Azaria in 1980," *Australia News*, www.guardian.co.uk/world/2012/jun/12/dingo-took-baby-azaria-chamberlain (accessed August 20, 2012).

18. C. N. Johnson and B. W. Brook, "Reconstructing the Dynamics of Ancient Human Populations from Radiocarbon Dates: 10 000 Years of Population Growth in Australia," *Proceedings of the Royal Society B* 278 (2011): 3748–54.

19. T. A. A. Prowse, C. N. J. Johnson, C. J. A. Bradshaw, and B. W. Brook, "An Ecological Regime Shift Resulting from Disrupted Predator-Prey Interactions in Holocene Australia," *Ecology* 95 (2014): 693–702.

20. A. D. Wallach, C. N. Johnson, E. G. Ritchie, and A. J. O'Neill, "Predator Control Promotes Invasive Dominated Ecological States," *Ecology Letters* 13 (2010): 1008–18.

21. D. G., Nimmo, D. Forsyth, S. J. Watson, and C. J. A. Bradshaw, "Dingoes Can Conserve Wildlife and Practitioners Can Tell," *Journal of Applied Ecology* 52 (2015): 283–85.

22. C. N. Johnson, J. L. Isaac, and D. O. Fisher, "Rarity of a Top Predator Triggers

Continent-Wide Collapse of Mammal Prey: Dingoes and Marsupials in Australia," *Proceedings of the Royal Society B* 274 (2007): 341–46.

23. M. Letnic, F. Koch, C. Gordon, M. S. Crowther, and C. R. Dickman, "Keystone Effects of an Alien Top-Predator Stem Extinctions of Native Mammals," *Proceedings of the Royal Society B* 276 (2009): 3249–56.

24. M. Letnic, E. G. Ritchie, and C. R. Dickman, "Top Predators as Biodiversity Regulators: The Dingo *Canis lupus dingo* as a Case Study," *Biological Reviews* 87 (2012): 390–413.

25. T. A. A. Prowse, C. N. Johnson, P. Cassey, C. J. A. Bradshaw, and B. W. Brook, "Ecological and Economic Benefits to Cattle Rangelands of Restoring an Apex Predator," *Journal of Applied Ecology* 52 (2015): 455–66.

26. Australian Collaborative Rangeland Information System. *Rangelands: Tracking Changes* (Canberra: Australian Collaborative Rangeland Information System, Commonwealth of Australia, 2001).

27. "Australia's Most Powerful Climate-Forcing Agent—It's Not Coal," *Brave New Climate*, August 11, 2008, bravenewclimate.com/2008/08/11/australias-most-powerful-climate-forcing-agent-its-not-coal/ (accessed July 9, 2014).

28. C. J. A. Bradshaw et al., "Brave New Green World—Consequences of a Carbon Economy for the Conservation of Australian Biodiversity," *Biological Conservation* 161 (2013): 71–90.

29. G. R. Wilson and M. J. Edwards, "Native Wildlife on Rangelands to Minimize Methane and Produce Lower-Emission Meat: Kangaroos versus Livestock," *Conservation Letters* 1 (2008): 119–28.

30. Ibid.

31. Australian Bureau of Statistics, www.abs.gov.au/AUSSTATS/abs@.nsf/mf/1383.0.55.001 (accessed August 8, 2014).

32. IUCN Red List of Threatened Species, www.iucnredlist.org (accessed August 20, 2012).

33. S. Blaber et al., "Elasmobranchs in Southern Indonesian Fisheries: The Fisheries, the Status of the Stocks and Management Options," *Reviews in Fish Biology and Fisheries* 19 (2009): 367–91.

34. I. C. Field, M. G. Meekan, R. C. Buckworth, and C. J. A. Bradshaw, "Protein Mining the World's Oceans: Australasia as an Example of Illegal Expansion-and-Displacement Fishing," *Fish and Fisheries* 10 (2009): 323–28.

35. S. Elks, "UN's Warning to Protect Barrier Reef," *The Australian*, June 2, 2012, www.theaustralian.com.au/national-affairs/uns-warning-to-protect-barrier-reef/story-fn59niix-1226381210119 (accessed August 20, 2012).

36. S. L. Pimm and R. A. Askins, "Forest Losses Predict Bird Extinctions in Eastern North America," *Proceedings of the National Academy of Sciences of the USA* 92 (1995): 9343–47.

37. D. R. Foster, "Land-Use History (1730–1990) and Vegetation Dynamics in Central New England, USA," *Journal of Ecology* 80 (1992): 753–71.

38. Pimm and Askins, "Forest Losses Predict Bird Extinctions in Eastern North America."

39. Ibid.

40. H. John Heinz III Center, *The State of the Nation's Ecosystems 2008* (Washington, DC: H. John Heinz Center for Science, Economics and the Environment, 2008).

41. M. Bürgi, E. W. B. Russell, and G. Motzkin, "Effects of Postsettlement Human Activities on Forest Composition in the North-Eastern United States: A Comparative Approach," *Journal of Biogeography* 27 (2000): 1123–38.

42. K. M. Flinn and M. Vellend, "Recovery of Forest Plant Communities in Post-Agricultural Landscapes," *Frontiers in Ecology and the Environment* 3 (2005): 243–50.

43. P. A. Martin, A. C. Newton, and J. M. Bullock, "Carbon Pools Recover More Quickly than Plant Biodiversity in Tropical Secondary Forests," *Proceedings of the Royal Society B: Biological Sciences* 280, 20132236 (2013).

44. J. F. Dwyer, D. J. Nowak, M. H. Noble, and S. M. Sisinni, "Connecting People with Ecosystems in the 21st Century: An Assessment of Our Nation's Urban Forests," *General Technical Report*, 1–483 (Evanston, IL: Pacific Northwest Research Station, USDA Forest Service, 2000).

45. F. Samson and F. Knopf, "Prairie Conservation in North America," *BioScience* 44 (1994): 418–21.

46. H. John Heinz III Center, *The State of the Nation's Ecosystems 2008.*

47. IUCN Red List of Threatened Species, www.iucnredlist.org.

48. J. B. C. Harris et al., "Conserving Imperiled Species: A Comparison of the IUCN Red List and U.S. Endangered Species Act," *Conservation Letters* 5 (2012): 64–72.

49. A. Prud'homme, *The Ripple Effect* (New York: Scribner, 2011).

50. N. M. Burkhead, "Extinction Rates in North American Freshwater Fishes, 1900–2010," *BioScience* 62 (2012): 798–808.

51. H. John Heinz III Center, *The State of the Nation's Ecosystems 2008.*

52. T. Gleeson, Y. Wada, M. F. P. Bierkens, and L. P. H. van Beek, "Water Balance of Global Aquifers Revealed by Groundwater Footprint," *Nature* 488 (2012): 197–200.

53. E. Stokstad, "Will Fracking Put Too Much Fizz in Your Water?" *Science* 344 (2014): 1468–71.

54. N. Oreskes, "A Green Bridge to Hell," *TomDispatch*, July 27, 2014, www.tomdispatch.com/blog/175873 (accessed August 8, 2014).

55. R. A. Myers, J. K. Baum, T. D. Shepherd, S. P. Powers, and C. H. Peterson, "Cascading Effects of the Loss of Apex Predatory Sharks from a Coastal Ocean," *Science* 315 (2007): 1846–50.

56. AP, "Warm Waters Sending Salmon to Canada, Not Washington," *Washington Times*, August 28, 2014, www.washingtontimes.com/news/2014/aug/28/warm-waters-sending-salmon-to-canada-not-wash/ (accessed September 16, 2014).

57. J. Hansen, M. Sato, and R. Ruedy, "Perception of Climate Change," *Proceedings of the National Academy of Sciences of the USA* 109 (2012): E2415–23.

58. D. Archer and S. Rahmstorf, *The Climate Crisis: An Introductory Guide to Climate Change* (New York: Cambridge University Press, 2010); N. Oreskes and E. M. Conway, *Merchants of Doubt: How a Handful of Scientists Obscured the Truth on Issues from Tobacco Smoke to Global Warming* (New York: Bloomsbury Press, 2010); J. Han-

sen, *Storms of My Grandchildren: The Truth about the Coming Climate Catastrophe and Our Last Chance to Save Humanity* (New York: Bloomsbury, 2009); M. E. Mann and L. R. Krump, *Dire Predictions: Understanding Global Warming: The Illustrated Guide to the Findings of the IPCC* (New York: DK Publishing, 2009); A. D. Barnosky, *Heatstroke* (Washington, DC: Island Press, 2009).

59. Oreskes and Conway, *Merchants of Doubt.*

60. C. John et al., "Quantifying the Consensus on Anthropogenic Global Warming in the Scientific Literature," *Environmental Research Letters* 8, 024024 (2013).

61. Intergovernmental Panel on Climate Change (IPCC), www.ipcc.ch (accessed August 20, 2012).

62. International Energy Organization (IEA), www.iea.org/stats (accessed August 20, 2012).

63. Garnet Climate Change Review, www.garnautreview.org.au (accessed August 20, 2012).

64. B. Parris, "Expanding Coal Exports Is Bad News for Australia and the World," *The Conversation*, September 12, 2013, theconversation.com/expanding-coal-exports-is-bad-news-for-australia-and-the-world-17937 (accessed July 21, 2014).

65. A. Y. Hoekstra and A. K. Chapagain, "Water Footprints of Nations: Water Use by People as a Function of Their Consumption Pattern," *Water Resources Management* 21 (2007): 35–48.

66. Ibid.

67. W. Gerbens-Leenes, A. Y. Hoekstra, and T. H. van der Meer, "The Water Footprint of Bioenergy," *Proceedings of the National Academy of Sciences of the USA* 106 (2009): 10219–23.

68. Hoekstra and Chapagain, "Water Footprints of Nations."

69. C. Jie, C. Jing-zhang, T. Man-zhi, and G. Zi-tong, "Soil Degradation: A Global Problem Endangering Sustainable Development," *Journal of Geographical Sciences* 12 (2002): 243–52.

70. A. G. J. Tacon and M. Metian, "Global Overview on the Use of Fish Meal and Fish Oil in Industrially Compounded Aquafeeds: Trends and Future Prospects," *Aquaculture* 285 (2008): 146–58.

CHAPTER SIX

1. S. Hayes, *Radical Homemakers. Reclaiming Domesticity from a Consumer Culture* (Richmondville, NY: Left to Write Press, 2010).

2. G. C. Daily, *Nature's Services* (Washington, DC: Island Press, 1997).

3. P. J. Crutzen, "Geology of Mankind: The Anthropocene," *Nature* 415 (2002): 23.

4. World Health Organization (WHO), "Global Burden of Disease," www.who.int/topics/global_burden_of_disease/en (accessed August 20, 2012).

5. A. J. McMichael, A. Nyong, and C. Corvalan, "Global Environmental Change and Health: Impacts, Inequalities, and the Health Sector," *British Medical Journal* 336 (2008): 191–94.

6. Ibid.

7. A. Costello et al., "Managing the Health Effects of Climate Change," *The Lancet* 373 (2009): 1693–733.

8. G. Wolff et al., *The World's Water, 2006–2007: The Biennial Report on Freshwater Resources* (Washington, DC: Island Press, 2006).

9. D. Pauly, R. Watson, and J. Alder, "Global Trends in World Fisheries: Impacts on Marine Ecosystems and Food Security," *Philosophical Transactions of the Royal Society B* 360 (2005): 5–12.

10. McMichael, Nyong, and Corvalan, "Global Environmental Change and Health."

11. A. Y. Vittor et al., "The Effect of Deforestation on the Human-Biting Rate of *Anopheles darlingi*, the Primary Vector of Falciparum Malaria in the Peruvian Amazon," *American Journal of Tropical Medicine and Hygiene* 74 (2006): 3–11; J. Yasuoka and R. Levins, "Impact of Deforestation and Agricultural Development on Anopheline Ecology and Malaria Epidemiology," *American Journal of Tropical Medicine and Hygiene* 76 (2007): 450–60.

12. C. J. A. Bradshaw, N. S. Sodhi, K. S. H. Peh, and B. W. Brook, "Global Evidence That Deforestation Amplifies Flood Risk and Severity in the Developing World," *Global Change Biology* 13 (2007): 2379–95.

13. L. C. Ivers and E. T. Ryan, "Infectious Diseases of Severe Weather-Related and Flood-Related Natural Disasters," *Current Opinion in Infectious Diseases* 19 (2006): 408–14.

14. K. E. Jones et al., "Global Trends in Emerging Infectious Diseases," *Nature* 451 (2008): 990–93.

15. Ibid.

16. A. Kallet and F. J. Schlink, *100,000,000 Guinea Pigs: Dangers in Everyday Foods, Drugs, and Cosmetics* (New York: Vanguard Press, 1933).

17. B. Cambreleng, "Fear Factor: China's Toxic Overload," *Agence France-Presse*, May 4, 2012, http://www.afp.com/en/node/96497 (accessed February 13, 2015).

18. J. Y. Suh, G. F. Birch, K. Hughes, and C. Matthai, "Spatial Distribution and Source of Heavy Metals in Reclaimed Lands of Homebush Bay: The Venue of the 2000 Olympic Games, Sydney, New South Wales," *Australian Journal of Earth Sciences* 51 (2004): 53–67.

19. G. F. Birch, B. Eyre, and S. E. Taylor, "The Distribution of Nutrients in Bottom Sediments of Port Jackson (Sydney Harbour), Australia," *Marine Pollution Bulletin* 38 (1999): 1247–51; S. McCready, D. J. Slee, G. F. Birch, and S. E. Taylor, "The Distribution of Polycyclic Aromatic Hydrocarbons in Surficial Sediments of Sydney Harbour, Australia," *Marine Pollution Bulletin* 40 (2000): 999–1006.

20. H. Bloom and G. Ayling, "Heavy Metals in the Derwent Estuary," *Environmental Geology* 2 (1977): 3–22.

21. P. S. Lake, D. Coleman, B. Mills, and R. Norris, "A Reconnaissance of Pollution of the King River in the Comstock-Crotty Area, West Tasmania," *Proceedings of the Royal Society of Tasmania* 1977 (1977): 157–73.

22. T. Colborn, D. Dumanoski, and J. P. Myers, *Our Stolen Future* (New York: Dutton, 1996).

23. The Endocrine Society. *Endocrine-Disrupting Chemicals: An Endocrine Society Scientific Statement* (2009), http://www.endocrine.org/~/media/endosociety/Files/Publications/Scientific%20Statements/EDC_Scientific_Statement.pdf (accessed July 29, 2014).

24. R. R. Newbold, E. Padilla-Banks, R. J. Snyder, and W. N. Jefferson, "Developmental Exposure to Estrogenic Compounds and Obesity," *Birth Defects Research (Part A)* 73 (2005): 478–80.

25. V. Drèze, G. Monod, J.-P. Cravedi, S. Biagianti-Risbourg, and F. Le Gac, "Effects of 4-Nonylphenol on Sex Differentiation and Puberty in Mosquitofish (*Gambusia holbrooki*)," *Ecotoxicology* 9 (2000): 93–103.

26. J. C. Semenza, P. E. Tolbert, C. H. Rubin, L. J. Guilette Jr., and R. J. Jackson, "Reproductive Toxins and Alligator Abnormalities at Lake Apopka, Florida," *Environmental Health Perspectives* 105 (1997): 1030–32.

27. C. A. Mackenzie, A. Lockridge, and M. Keith, "Declining Sex Ratio in a First Nation Community," *Environmental Health Perspectives* 113 (2005): 1295–98.

28. Ibid.

29. D. L. Davis et al., "Declines in Sex Ratio at Birth and Fetal Deaths in Japan, and in U.S. Whites but Not African Americans," *Environmental Health Perspectives* 115 (2007): 941–46.

30. L. N. Vandenberg et al., "Hormones and Endocrine-Disrupting Chemicals: Low-Dose Effects and Nonmonotonic Dose Responses," *Endocrine Reviews* 33 (2012): 378–455.

31. Kallet and Schlink, *100,000,000 Guinea Pigs.*

32. FDA, "Bisphenol A (BPA): Use in Food Contact Application," www.fda.gov/newsevents/publichealthfocus/ucm064437.htm (accessed August 23, 2012).

33. F. R. Kaufman, "Type 2 Diabetes in Children and Young Adults: A 'New Epidemic.'" *Clinical Diabetes* 20 (2002): 217–18.

34. K. Grieger, A. Baun, and R. Owen, "Redefining Risk Research Priorities for Nanomaterials," *Journal of Nanoparticle Research* 12 (2010): 383–92.

35. J. Cribb, *Poisoned Planet* (Sydney: Allen & Unwin, 2014).

36. S. Barrett et al., "Climate Engineering Reconsidered," *Nature Climate Change* 4 (2014): 527–29.

37. Safe Planet, "5 Millions Deaths Annual Are Due to Chemical Exposure: Many of These Deaths Are Preventable," www.safepla.net/exp-who.html (accessed July 26, 2014).

38. Cribb, *Poisoned Planet.*

39. T. McMichael, "Book Review: *Poisoned Planet*," *The Conversation*, June 24, 2014, theconversation.com/book-review-poisoned-planet-28255 (accessed July 26, 2014).

40. E. Malaj et al., "Organic Chemicals Jeopardize the Health of Freshwater Ecosystems on the Continental Scale," *Proceedings of the National Academy of Sciences of the USA* 111 (2014): 9549–54.

41. C. J. A. Bradshaw and B. W. Brook, "Human Population Reduction Is Not a Quick Fix for Environmental Problems," *Proceedings of the National Academy of Sciences of the USA* 111 (2014): 16610–15.

42. J. A. Tainter, *The Collapse of Complex Societies* (Cambridge: Cambridge University Press, 1988).

43. M. T. Klare, *The Race for What's Left: The Global Scramble for the World's Last Resources* (New York: Metropolitan Books, 2012).

44. Bradshaw and Brook, "Human Population Reduction Is Not a Quick Fix for Environmental Problems."

45. S. G. Potts et al., "Global Pollinator Declines: Trends, Impacts and Drivers," *Trends in Ecology and Evolution* 25 (2010): 345–53.

46. Syngenta, www.syngenta.com (accessed August 20, 2012).

47. N. Oreskes and E. M. Conway, *Merchants of Doubt: How a Handful of Scientists Obscured the Truth on Issues from Tobacco Smoke to Global Warming* (New York: Bloomsbury Press, 2010).

48. Center for Media and Democracy, "Steven J. Milloy," *Sourcewatch*, www.sourcewatch.org/index.php?title=Steven_J._Milloy (accessed August 20, 2012).

49. A. Hess, "Fifty Years After *Silent Spring*, Sexism Persists in Science," *Slate*, September 26, 2012, www.slate.com/blogs/xx_factor/2012/09/28/_50_years_after_rachel_carson_s_silent_spring_sexism_persists_in_science_.html (accessed September 16, 2014).

50. Center for Environmental Health (CEH), "When Corporations Attack," podcast, July 29, 2013, www.ceh.org/news-events/podcasts/content/when-corporations-attack/ (accessed September 16, 2014).

CHAPTER SEVEN

1. A. H. Westing, "All the Many Humans Ever: An Update," *BioScience* 60 (2010): 777.

2. S. Singh, G. Sedgh, and R. Hussain, "Unintended Pregnancy: Worldwide Levels, Trends, and Outcomes," *Studies in Family Planning* 41 (2010): 241–50.

3. Ibid.

4. P. R. Ehrlich, "The MAHB, the Culture Gap, and Some Really Inconvenient Truths," *PLoS Biology* 8 (2010): e1000330.

5. P. R. Ehrlich and J. Holdren, "Impact of Population Growth," *Science* 171 (1971): 1212–17.

6. J. P. Holdren and P. R. Ehrlich, "Human Population and the Global Environment," *American Scientist* 62 (1974): 282–92.

7. B. W. Brook and C. J. A. Bradshaw, "Strength of Evidence for Density Dependence in Abundance Time Series of 1198 Species," *Ecology* 87 (2006): 1445–51; S. Herrando-Pérez, S. Delean, B. W. Brook, and C. J. A. Bradshaw, "Density Dependence: An Ecological Tower of Babel," *Oecologia* 170 (2012): 585–603.

8. Herrando-Pérez et al., "Density Dependence."

9. T. Flannery, in *Australian Academy of Science Annual Symposium: Population 2040: Australia's Choice* (Australian Academy of Science), 47–61.

10. O. L. I. Brown, A. Hammill, and R. McLeman, "Climate Change as the 'New' Security Threat: Implications for Africa," *International Affairs* 83 (2007): 1141–54.

11. S. Solomon, G.-K. Plattner, R. Knutti, and P. Friedlingstein, "Irreversible Climate Change Due to Carbon Dioxide Emissions," *Proceedings of the National Academy of Sciences of the USA* 106 (2009): 1704–9.

12. H. K. Gibbs et al., "Tropical Forests Were the Primary Sources of New Agricultural Land in the 1980s and 1990s," *Proceedings of the National Academy of Sciences of the USA* 107 (2010): 16732–37.

13. G. C. Daily, *Nature's Services* (Washington, DC: Island Press, 1997).

14. S. L. Pimm et al., "The Biodiversity of Species and Their Rates of Extinction, Distribution, and Protection," *Science* 344, doi:10.1126/science.1246752 (2014).

15. Authors Place, "Fred Pearce," authorsplace.co.uk/fred-pearce (accessed August 20, 2012).

16. C. J. A. Bradshaw, X. Giam, and N. S. Sodhi, "Evaluating the Relative Environmental Impact of Countries," *PLoS One* 5 (2010): e10440.

17. C. J. A. Bradshaw and B. W. Brook, "Human Population Reduction Is Not a Quick Fix for Environmental Problems," *Proceedings of the National Academy of Sciences of the USA* 111 (2014): 16610–15.

18. Ehrlich and Holdren, "Impact of Population Growth"; J. Harte, "Human Population as a Dynamic Factor in Environmental Degradation," *Population and Environment* 28 (2007): 223–36.

19. J. K. Levy and C. Gopalakrishnan, "Promoting Ecological Sustainability and Community Resilience in the US Gulf Coast After the 2010 Deepwater Horizon Oil Spill," *Journal of Natural Resources Policy Research* 2 (2010): 297–315.

20. B. R. Silliman et al., "Degradation and Resilience in Louisiana Salt Marshes After the BP–Deepwater Horizon Oil Spill," *Proceedings of the National Academy of Sciences of the USA* 109 (2012): 11234–39; H. K. White et al., "Impact of the Deepwater Horizon Oil Spill on a Deep-Water Coral Community in the Gulf of Mexico," *Proceedings of the National Academy of Sciences of the USA* 109 (2012): 20303–8.

21. J. Lorenz, "BP to Try Deepwater Drilling in Australia," *ABC: Environment*, September 26, 2011, www.abc.net.au/environment/articles/2011/09/26/3324770.htm (accessed August 20, 2012).

22. L. N. Vandenberg Theo Colborn, Tyrone B. Hayes, Jerrold J. Heindel, David R. Jacobs, Jr., Duk-Hee Lee, Toshi Shioda, Ana M. Soto, Frederick S. vom Saal, Wade V. Welshons, R. Thomas Zoeller, John Peterson Myers. "Hormones and Endocrine-Disrupting Chemicals: Low-Dose Effects and Nonmonotonic Dose Responses," *Endocrine Reviews* 33 (2012): 378–455.

23. G. C. Daily and P. R. Ehrlich, "Impacts of Development and Global Change on the Epidemiological Environment," *Environment and Development Economics* 1 (1996): 309–44.

24. P. R. Ehrlich, *The Population Bomb* (New York: Ballantine Books, 1968).

25. Oak Ridge National Laboratory, *Nuclear Energy Centers, Industrial and Agro-industrial Complexes* (Summary Report, ORNL-4291, 1968).

26. S. Lane et al., "Australians Wary of 36m Population," *ABC: News*, April 7, 2010, www.abc.net.au/news/2010-04-08/australians-wary-of-36m-population-target/2584738 (accessed August 20, 2012).

27. W. E. Rees, in *Encyclopedia of Biodiversity*, vol. 2, ed. S. A. Levin (San Diego, CA: Academic Press, 2001), 229–44.

28. Global Footprint Network, www.footprintnetwork.org/en/index.php/GFN (accessed August 20, 2012).

29. D. Brooks, "The Opinion Pages," *New York Times*, brooks.blogs.nytimes.com (accessed August 20, 2012).

30. D. Brooks, "The Fertility Implosion," *New York Times*, March 13, 2012, A25.

31. P. R. Ehrlich and A. H. Ehrlich, "Enough Already," *New Scientist* 191 (2006): 46–50.

32. J. Bongaarts, "Human Population Growth and the Demographic Transition," *Philosophical Transactions of the Royal Society B: Biological Sciences* 364 (2009): 2985–90.

33. Bradshaw and Brook, "Human Population Reduction Is Not a Quick Fix for Environmental Problems."

34. D. E. Bloom, D. Canning, and G. Fink, "Implications of Population Ageing for Economic Growth," *Oxford Review of Economic Policy* 26 (2010): 583–612.

35. B. Hartmann, www.betsyhartmann.com (accessed August 20, 2012).

36. B. Hartmann, "The 'New' Population Control Craze: Retro, Racist, Wrong Way to Go," *Issues Magazine* (Fall 2009).

37. The Greens, greens.org.au (accessed August 20, 2012).

38. D. T. Blumstein and C. Saylan, *The Failure of Environmental Education (And How We Can Fix It)* (Berkeley: University of California Press, 2011); P. R. Ehrlich, "A Personal View: Environmental Education—Its Content and Delivery," *Journal of Environmental Studies and Sciences* 1 (2011): 6–13.

39. P. R. Ehrlich, *Human Natures: Genes, Cultures, and the Human Prospect* (Washington, DC: Island Press, 2000).

40. Food and Agriculture Organization of the United Nations, *The State of Food Insecurity in the World: Addressing Food Insecurity in Protracted Crises* (Food and Agriculture Organization of the United Nations, Rome, 2010), www.fao.org/docrep/013/i1683e/i1683e.pdf (accessed February 13, 2015).

41. Bradshaw and Brook, "Human Population Reduction Is Not a Quick Fix for Environmental Problems."

42. D. J. Davidson, J. Andrews, and D. Pauly, "The Effort Factor: Evaluating the Increasing Marginal Impact of Resource Extraction Over Time," *Global Environmental Change* 25 (2014): 63–68.

CHAPTER EIGHT

1. "Rinehart's Welfare Comments an 'Insult to Millions': Treasurer," *Sydney Morning Herald*, August 30, 2012, www.smh.com.au/opinion/political-news/rineharts-welfare-comments-an-insult-to-millions-treasurer-20120830-251u5.html (accessed August 30, 2012).

2. C. Mooney, *The Republican War on Science* (New York: Basic Books, 2006); C. Mooney, *Unscientific America: How Scientific Illiteracy Threatens our Future* (New

York: Basic Books, 2009); C. Mooney, *The Republican Brain: The Science of Why They Deny Science and Reality* (New York: Wiley, 2012).

3. Science Blogs, “The War on Science,” scienceblogs.com/deltoid/category/the_war_on_science (accessed August 20, 2012).

4. Mooney, *The Republican War on Science.*

5. N. Oreskes and E. M. Conway, *Merchants of Doubt: How a Handful of Scientists Obscured the Truth on Issues from Tobacco Smoke to Global Warming* (New York: Bloomsbury Press, 2010).

6. Heartland Institute, heartland.org (accessed August 20, 2012).

7. A. Revkin, “The Short Hot Life of Heartland’s Hateful Climate Billboard,” *New York Times*, May 4, 2012, dotearth.blogs.nytimes.com/2012/05/04/the-short-hot-life-of-heartlands-hateful-climate-billboard (accessed August 20, 2012).

8. ITS Global, www.itsglobal.net (accessed August 20, 2012).

9. World Growth, www.worldgrowth.org (accessed August 20, 2012).

10. “Wolves in Sheep’s Clothing: Industrial Lobbyists and the Destruction of Tropical Forests,” *ConservationsBytes.com*, October 25, 2010, conservationbytes.com/2010/10/25/wolves-sheeps-clothing (accessed August 20, 2012).

11. W. F. Laurance et al., “Predatory Corporations, Failing Governance, and the Fate of Forests in Papua New Guinea,” *Conservation Letters* 4 (2011): 95–100; M. Colchester, *Palm Oil and Indigenous Peoples in Southeast Asia: Land Acquisition, Human Rights Violations and Indigenous Peoples on the Palm Oil Frontier* (Moreton-in-Marsh, UK: Forest Peoples Programme, 2010).

12. W. L. Robinson, *Global Capitalism and the Crisis of Humanity* (New York: Cambridge University Press, 2014).

13. *Growth Busters*, www.growthbusters.org (accessed August 20, 2012).

14. postgrowth.org (accessed 12 October 2012).

15. Millennium Alliance for Humanity and the Biosphere (MAHB), mahb.stanford.edu (accessed August 20, 2012).

16. S. Santow, “Science Literacy at Risk of Extinction,” *ABC: News*, August 31, 2010, www.abc.net.au/news/2010-07-30/science-literacy-at-risk-of-extinction/925484 (accessed August 20, 2012); “How Long Does It Take the Earth to Go Around the Sun?,” youtu.be/cU2dZz18P0c (accessed August 20, 2012).

17. “Fox News Viewers Know Less Than People Who Don’t Watch Any News: Study,” *Huffington Post*, November 21, 2011, www.huffingtonpost.com/2011/11/21/fox-news-viewers-less-informed-people-fairleigh-dickinson_n_1106305.html (accessed November 22, 2011).

18. C. Berg, “Phoney Food Fears Ignore Nimble Market Solutions,” *Sydney Morning Herald*, December 4, 2011, www.smh.com.au/opinion/society-and-culture/phoney-food-fears-ignore-nimble-market-solutions-20111203-1ocl4.html (accessed December 4, 2011).

19. *Crikey*, crikey.com.au (accessed March 16, 2015).

20. The Conversation, theconversation.edu.au (accessed August 20, 2012).

21. *New Matilda*, http://newmatilda.com (accessed October 25, 2014).

22. B. Lomborg, *The Skeptical Environmentalist: Measuring the Real State of the World* (Cambridge: Cambridge University Press, 2001).

23. Access to some scientific reviews can be found at the website of the Union of Concerned Scientists (www.ucsusa.org) (accessed August 24, 2012).

24. For a complete debunking by many experts, see K. Lambeck, "Debunking Ian Plimer's 'Heaven and Earth,'" *To Be Precise*, tbp.mattandrews.id.au/2009/06/06/debunking-plimer-heaven-and-earth (accessed August 24, 2012).

25. Union of Concerned Scientists, www.ucsusa.org.

26. G. Readfearn, "The Millions Behind Bjorn Lomborg's Copenhagen Consensus Center US Think Tank," *Desmog Blog*, June 24, 2014, www.desmogblog.com/2014/06/25/millions-behind-bjorn-lomborg-copenhagen-consensus-center (accessed August 3, 2014).

27. M. T. Klare, *The Race for What's Left: The Global Scramble for the World's Last Resources* (New York: Metropolitan Books, 2012).

28. B. Freeman, "Navy Official Discusses Climate Change Investment Strategy," US Department of Defense, June 21, 2010, www.defense.gov/news/newsarticle.aspx?id=59713 (accessed August 24, 2012).

29. N. Diamond-Smith and M. Potts, "Are the Population Policies of India and China Responsible for the Fertility Decline?" *International Journal of Environmental Studies* 67 (2010): 291–301.

30. P. R. Ehrlich and A. H. Ehrlich, "Enough Already," *New Scientist* 191 (2006): 46–50.

31. Bradshaw and Brook, "Human Population Reduction Is Not a Quick Fix for Environmental Problems."

32. I. S. Fish, "Lost in America: How America's Favorite Chinese Dissident Burned Bridges at NYU and Ended Up at a Pro-Life, Anti-Gay Think Tank," *Foreign Policy*, January 10, 2014, www.foreignpolicy.com/articles/2014/01/10/lost_in_america_chen_guangcheng (accessed July 29, 2014).

33. Bradshaw and Brook, "Human Population Reduction Is Not a Quick Fix for Environmental Problems."

34. "Death Tolls Across History," *Necrometrics*, http://necrometrics.com (accessed July 30, 2014).

35. Bradshaw and Brook, "Human Population Reduction Is Not a Quick Fix for Environmental Problems."

36. Ibid.

37. Ibid.

38. B. Lomborg, "Environmental Alarmism, Then and Now," *Foreign Affairs* 91 (2012): 24–40.

39. P. R. Ehrlich, "A Personal View: Environmental Education—Its Content and Delivery," *Journal of Environmental Studies and Sciences* 1 (2011): 6–13.

40. Oreskes and Conway, *Merchants of Doubt*; R. E. Dunlap and A. M. McCright, in *Routledge Handbook of Climate Change and Society*, ed. C. Lever-Tracey (Milton Park, UK: Taylor & Francis, 2010), 240–60.

41. P. R. Ehrlich, in *Ethics in Science and Environmental Ethics,* ed. D. Pauly (Oldendorf/Luhe, Germany: Inter-Research, 2015).

42. I. Kubiszewski et al., "Beyond GDP: Measuring and Achieving Global Genuine Progress," *Ecological Economics* 93 (2013): 57–68.

43. D. C. Korten, *Agenda for a New Economy: From Phantom Wealth to Real Wealth* (San Francisco: Berrett-Koehler Publishers, 2009).

44. "The Changing Wealth of Nations," *World Bank,* data.worldbank.org/data-catalog/wealth-of-nations (accessed August 20, 2012).

45. Kubiszewski et al., "Beyond GDP."

46. Ibid.

47. OpenSecrets.org: Center for Responsive Politics, www.opensecrets.org (accessed August 20, 2012).

48. T. Hamburger and W. Heisel, "Heavily Invested in the Outcome," *Los Angeles Times,* September 24, 2008, articles.latimes.com/2008/sep/23/business/fi-polmoney23 (accessed June 3, 2011).

49. M. Vasilogambros and S. Mimms, "Scalia Defends Citizens United Decision," *National Journal,* July 18, 2012, www.nationaljournal.com/politics/scalia-defends-citizens-united-decision-reflects-on-term-in-rare-tv-appearance-20120718 (accessed August 20, 2012).

50. A. Kroll, "The Supreme Court Just Gutted Another Campaign Finance Law," *Mother Jones,* April 2, 2014, www.motherjones.com/politics/2014/03/supreme-court-mccutcheon-citizens-united (accessed August 3, 2014).

51. J. K. Galbraith, *The Predator State: How Conservatives Abandoned the Free Market and Why Liberals Should Too* (New York: Free Press, 2008).

52. M. Gallucci, "Obama Paves the Way for East Coast Offshore Oil Exploration with Controversial Sonic Cannons," *International Business Times,* July 18, 2014, www.ibtimes.com/obama-paves-way-east-coast-offshore-oil-exploration-controversial-sonic-cannons-1632726 (accessed August 3, 2014).

53. "Jim Inhofe," *Dirty Energy Money,* dirtyenergymoney.com/view.php?searchvalue=inhofe&search=1&type=search&searchtype=can#view=connections (accessed September 30, 2012); "Jim Inhofe," *Republican Retard Club,* July 2011, republicanretardclub.blogspot.com.au/2011/07/jim-inhofe.html (accessed September 30, 2012).

54. A. Austin, "Tony Abbott Continues to Trash Australia's International Reputation Abroad," *Independent Australia,* July 9, 2014, www.independentaustralia.net/politics/politics-display/abbott-continues-to-trash-australias-international-reputation,6652 (accessed August 1, 2014).

55. K. Mathiesen, "UN Rejects Australia's 'Feeble' Bid to Strip Tasmanian Forest's Heritage Status," *Guardian,* June 24, 2014, www.theguardian.com/environment/2014/jun/23/un-rejects-australia-tasmanian-forest-heritage (accessed August 1, 2014).

56. O. Milman, "Great Barrier Reef Impact from Dredging Could Cost 'as Much as $1bn.'" *Guardian,* July 12, 2014, http://www.theguardian.com/environment/2014/jul/11/great-barrier-reef-impact-from-dredging-could-cost-as-much-as-1bn (accessed August 1, 2014).

57. E. G. Ritchie et al., "Continental-Scale Governance and the Hastening of Loss of Australia's Biodiversity," *Conservation Biology* 27 (2013): 1133–35.

58. "Scientists Blast Australia Leader's Proposed Ban on Parks," *Mongabay.com*, March 5, 2014, news.mongabay.com/2014/0305-abbott-ban-on-parks-alert.html (accessed August 1, 2014).

59. M. Khadem, "Bid to Strip Green Groups' Tax Status," *The Age*, August 9, 2006, www.theage.com.au/news/national/bid-to-strip-green-groups-tax-status/2006/08/08/1154802891527.html (accessed August 1, 2014).

60. L. Taylor, "Australian Government May Ban Environmental Boycotts," *Guardian*, April 2, 2014, www.theguardian.com/environment/2014/apr/02/coalition-review-of-consumer-laws-may-ban-environmental-boycotts (accessed August 1, 2014).

61. N. Towell, "Environment Department Cuts Staff in a Bid to Slash Budget," *Sydney Morning Herald*, April 7, 2014, www.smh.com.au/national/public-service/environment-department-cuts-staff-in-a-bid-to-slash-budget-20140407-368cv.html (accessed August 1, 2014).

62. T. Allan, "Landcare and Research Cuts in Budget," *ABC: Rural*, May 13, 2014, www.abc.net.au/news/2014-05-13/budget-overview/5441510 (accessed August 1, 2014).

63. D. Pannell, "Litany of Deep Cuts for Environmental Programs," *The Conversation*, May 14, 2014, http://theconversation.com/litany-of-deep-cuts-for-environmental-programs-26499 (accessed August 1, 2014).

64. G. McIntyre, "Australia's Environment Minister Could Soon Be Above the Law," *The Conversation*, Febuary 27, 2014, http://theconversation.com/australias-environment-minister-could-soon-be-above-the-law-23361 (accessed August 1, 2014).

65. R. Su, "Tony Abbott, Coalition Cut Environmental Defenders Offices Budget; Move Slammed as 'Barbaric,'" *Environmental Justice Australia*, December 19, 2013, http://envirojustice.org.au/media/tony-abbott-coalition-cut-environmental-defenders-offices-budget-move-slammed-as-barbaric (accessed August 1, 2014).

66. "Will Australia Back-Slide on Its Illegal Logging Bill?," *ALERT*, January 31, 2014, http://alert-conservation.org/issues-research-highlights/2014/1/31/will-australia-back-slide-on-its-illegal-logging-bill (accessed August 1, 2014).

67. L. Taylor, "Australia Kills Off Carbon Tax," *Guardian*, July 17, 2014, www.theguardian.com/environment/2014/jul/17/australia-kills-off-carbon-tax (accessed August 1, 2014).

68. G. Coslovich, "Memo Abbott: Virginity Debate Is No Man's Land," *The Age*, January 28, 2010, www.theage.com.au/opinion/memo-abbott-virginity-debate-is-no-mans-land-20100127-mz0y.html (accessed August 20, 2012).

69. "Julia Gillard's Misogny Speech," October 10, 2012, www.youtube.com/watch?v=SOPsxpMzYw4 (accessed August 1, 2014).

70. L. Wilson, "1000-Year Vision Fuels Climate Fight," *The Australian*, March 29, 2011, www.theaustralian.com.au/national-affairs/year-vision-fuels-climate-fight/story-fn59niix-1226029695904 (accessed August 20, 2012).

71. G. Jennett, "Abbott Pounces on Carbon Tax Admission," *ABC: News*, March 15, 2011, www.abc.net.au/news/2011-03-15/abbott-pounces-on-carbon-tax-admission/2656650 (accessed August 20, 2012).

72. L. Taylor, "Carbon Dioxide Not the Bad Guy, Says Abbott," *The Age*, March 15, 2011, www.theage.com.au/environment/climate-change/carbon-dioxide-not-the-bad-guy-says-abbott-20110314-1bul3.html (accessed August 20, 2012).

73. Lambeck, "Debunking Ian Plimer's 'Heaven and Earth.'"

74. M. Wilkinson, "Climate Sceptics Have Made Their Triumphant Return," *Sydney Morning Herald*, December 2, 2009, www.smh.com.au/opinion/politics/climate-sceptics-have-made-their-triumphant-return-20091201-k3xm.html (accessed August 20, 2012).

75. "Climate Misinformer: Tony Abbott," *Skeptical Science*, www.skepticalscience.com/skeptic_Tony_Abbott.htm (accessed August 20, 2012).

76. B. Packham and J. Massola, "Miners Spent $21m to Beat Resource Tax, Political Donations Figures Show," *The Australian*, February 1, 2011, www.theaustralian.com.au/national-affairs/miners-spent-21m-to-beat-resource-tax-political-donations-figures-show/story-fn59niix-1225997986153 (accessed August 20, 2012).

77. Donations can be viewed online at "Annual Returns Locator Service," *Australian Electoral Commission*, periodicdisclosures.aec.gov.au (accessed August 20, 2012).

78. J. West, "Mining Magnate, Property Tycoon—Politician? Just Who Is Clive Palmer?," *The Conversation*, April 30, 2012, http://theconversation.com/mining-magnate-property-tycoon-politician-just-who-is-clive-palmer-6646 (accessed August 2, 2014).

79. M. Grattan, "Clive Palmer Promises Carbon Tax Repeal in a Policy That Throws Bones in All Directions," *The Conversation*, June 25, 2014, http://theconversation.com/clive-palmer-promises-carbon-tax-repeal-in-a-policy-that-throws-bones-in-all-directions-28475 (accessed August 2, 2014).

80. H. Aston, "Clive Palmer Blames Tony Abbott for Falling Asleep in Parliament," *Sydney Morning Herald*, May 14, 2014, www.smh.com.au/federal-politics/political-news/clive-palmer-blames-tony-abbott-for-falling-asleep-in-parliament-20140514-zrcm8.html (accessed August 2, 2014).

81. *The Ellen Degeneres Show*, ellen.warnerbros.com (accessed August 20, 2012).

82. D. Colander and A. Klamer, "The Making of an Economist," *Journal of Economic Perspectives* 1 (1987): 95–111.

83. G. Chin, T. Marathe, and L. Roberts, "Population: Doom or Vroom," *Science* 333 (2011): 539.

84. W. I. Robinson, *Global Capitalism and the Crisis of Humanity* (New York: Cambridge University Press, 2014).

85. N. Klein, *This Changes Everything: Capitalism vs. the Climate* (New York: Simon & Schuster, 2014).

86. "Noam Chomsky capitalism" search, *Ecosia*, http://ecosia.org/search?q=noam+chomsky+capitalism&addon=opensearch#gsc.tab=0&gsc.q=noam%20chomsky%20capitalism&gsc.page=1 (accessed October 23, 2014).

87. "Gagging Scientists about Marine Parks," *ConservationBytes.com*, August 14, 2011, conservationbytes.com/2011/08/14/gagging-scientists-about-marine-parks (accessed August 20, 2012); "Bush Administration Accused of Gagging Scientists," *Tiny Tuna*, May 30, 2007, http://tinytuna.com/bush-administration-accused-of-gagging-scientists/ (accessed August 20, 2012); B. Johnson, "DOJ Gags Scientists Studying BP Disaster," *Think Progress*, August 10, 2010, thinkprogress.org/politics/2010/08/10/112945/scientists-bp-gag (accessed August 20, 2012).

CHAPTER NINE

1. M. Goldberg, "A Christian Plot for Domination?" *Daily Beast*, August 11, 2011, www.thedailybeast.com/articles/2011/08/14/dominionism-michele-bachmann-and-rick-perry-s-dangerous-religious-bond.html (accessed August 20, 2012).

2. Office of the Governor Greg Abbott, governor.state.tx.us/news/proclamation/16038 (accessed October 12, 2011).

3. E. McMorris-Santoro, "Rick Perry Climate Change Is a Hoax Drummed Up by Scientists Looking to Make Money," *TPM*, August 17, 2011, tpmdc.talkingpointsmemo.com/2011/08/rick-perry-climate-change-is-a-hoax-drummed-up-by-scientists-looking-to-make-money.php?ref=fpb (accessed September 29, 2011).

4. J. Cook et al., "Quantifying the Consensus on Anthropogenic Global Warming in the Scientific Literature," *Environmental Research Letters* 8 (2013): 024024.

5. "Michelle Bachmann on Families and Children," *On the Issues*, www.ontheissues.org/house/Michele_Bachmann_Families_+_Children.htm (accessed July 22, 2011).

6. K. Gushta, "Be Fruitful and Multiply," *Christian Post*, August 16, 2011, www.christianpost.com/news/be-fruitful-and-multiply-54042 (accessed November 22, 2011).

7. R. Quigley, "19 Kids and Counting Becomes 20: Duggar Parents Announce They Are Expecting Another Child," *Daily Mail*, November 9, 2011, www.dailymail.co.uk/news/article-2059016/Michelle-Duggar-pregnant-19-Kids-Counting-20.html (accessed November 22, 2011).

8. N. George, "Catholic Church in India Says Have More Children," *Boston.com*, October 11, 2011, www.boston.com/news/world/asia/articles/2011/10/11/catholic_church_in_india_says_have_more_children (accessed November 24, 2011).

9. T. W. Jones, "Bernardi Slips Down the Political Slope with Bestial Comments on Marriage," *The Conversation*, September 19, 2012, theconversation.edu.au/bernardi-slips-down-the-political-slope-with-bestial-comments-on-marriage-9685 (accessed September 30, 2012).

10. Family First, www.familyfirst.org.au (accessed August 20, 2012).

11. "Marriage Is Special," *Family First*, www.nsw-familyfirst.org.au/marriage (accessed August 20, 2012).

12. "Climate Change," *Family First*, www.nsw-familyfirst.org.au/climate-change (accessed August 20, 2012).

13. B. Westcott, "Religion in Schools: Why Teachers, Parents Are Powerless to Stop

It," *Crikey*, August 17, 2012, www.crikey.com.au/2012/08/17/religion-in-schools-why-teachers-parents-are-powerless-to-stop-it (accessed August 20, 2012).

14. S. Cory, "Judgement Day for Abbott on Science and Research Funding," *The Conversation*, May 15, 2014, http://theconversation.com/judgement-day-for-abbott-on-science-and-research-funding-26684 (accessed February 14, 2015).

15. P. Lewis, "Australia's Economy Is Healthy, So How Can There Be a Budget Crisis?" *The Conversation*, May 5, 2014, http://theconversation.com/australias-economy-is-healthy-so-how-can-there-be-a-budget-crisis-26036 (accessed August 2, 2014).

16. C. Chang, "High Court Rules Against Federal Government Funding of School Chaplains," *News.com.au*, June 19, 2014, www.news.com.au/national/high-court-rules-against-federal-government-funding-of-school-chaplains/story-fncynjr2-1226959661271 (accessed August 2, 2014).

17. Population Party, www.populationparty.org.au (accessed August 20, 2012).

18. L. D. Wilke, "Nuns Gone Rogue," *Huffington Post*, August 14, 2012, www.huffingtonpost.com/lorraine-devon-wilke/nuns-gone-rogue_b_1771575.html (accessed August 20, 2012).

19. A. Hasler, *How the Pope Became Infallible: Pius IX and the Politics of Persuasion* (New York: Doubleday, 1981); A. Weisman, *Countdown: Our Last, Best Hope for a Future on Earth?* (New York: Little, Brown, 2013).

20. G. Lakoff and E. Wehling, "The Sacredness of Life and Liberty," *Huffington Post*, July 16, 2012, www.huffingtonpost.com/george-lakoff/birth-control-framing_b_1675759.html (accessed August 20, 2012).

21. S. A. Bhagwat, N. Dudley, and S. R. Harrop, "Religious Following in Biodiversity Hotspots: Challenges and Opportunities for Conservation and Development," *Conservation Letters* 4 (2011): 234–40.

22. C. Hitchens, "Mommie Dearest," *Slate*, October 20, 2003, www.slate.com/articles/news_and_politics/fighting_words/2003/10/mommie_dearest.html (accessed November 24, 2011).

23. Bhagwat, Dudley, and Harrop, "Religious Following in Biodiversity Hotspots."

24. D. Frum, "How Rick Perry Got Rich," *Frum Forum*, August 30, 2011, www.frumforum.com/how-rick-perry-got-rich (accessed August 20, 2012).

25. M. Williams, "Republican Congressman Paul Broun Dismisses Evolution and Other Theories," *Guardian*, October 6, 2012, www.guardian.co.uk/world/2012/oct/06/republican-congressman-paul-broun-evolution-video (accessed October 10, 2012).

26. C. Hitchens, *God Is Not Great: How Religion Poisons Everything* (New York: Twelve, 2007).

27. P. R. Ehrlich and R. E. Ornstein, *Humanity on a Tightrope: Thoughts on Empathy, Family, and Big Changes for a Viable Future* (New York: Rowman & Littlefield, 2010).

28. P. R. Ehrlich, *Human Natures: Genes, Cultures, and the Human Prospect* (Washington, DC: Island Press, 2000).

29. G. S. Paul, "Cross-National Correlations of Quantifiable Societal Health with

Popular Religiosity and Secularism in the Prosperous Democracies: A First Look," *Journal of Religion and Society* 7 (2005): 1–17.

30. R. Dawkins, *The God Delusion* (New York: Houghton Mifflin, 2006).

31. C. Kimball, *When Religion Becomes Evil: Five Warning Signs* (San Francisco: Harper Collins, 2003).

CHAPTER TEN

1. F. Lyman, I. Mintzer, K. Courrier, and J. MacKenzie, *The Greenhouse Trap: What We're Doing to the Atmosphere and How We Can Slow Global Warming* (Boston: Beacon Press, 1990).

2. National Academy of Sciences USA, in *Population Summit of the World's Scientific Academies* (Washington, DC: National Academies Press, 1993); Union of Concerned Scientists, *World Scientists' Warning to Humanity* (Cambridge, MA: Union of Concerned Scientists, 1993).

3. C. J. A. Bradshaw and B. W. Brook, "Human Population Reduction Is Not a Quick Fix for Environmental Problems," *Proceedings of the National Academy of Sciences of the USA* 111 (2014): 16610–15.

4. W. E. Rees, in *Encyclopedia of Biodiversity*, 2nd ed., ed. S. Levin (San Diego, CA: Academic Press, 2013).

5. Global Footprint Network, www.footprintnetwork.org/en/index.php/GFN (accessed August 20, 2012).

6. D. J. Davidson, J. Andrews, and D. Pauly, "The Effort Factor: Evaluating the Increasing Marginal Impact of Resource Extraction Over Time," *Global Environmental Change* 25 (2014): 63–68; D. J. Davidson and J. Andrews, "Not All about Consumption," *Science* 339 (2013): 1286–87.

7. O. Toon et al., "Consequences of Regional-Scale Nuclear Conflicts," *Science* 315 (2007): 1224–25.

8. D. Coumou and S. Rahmstorf, "A Decade of Weather Extremes," *Nature Climate Change* 2 (2012): 491–96; Climate Council of Australia, *Angry Summer 2013/2014* (Sydney: Climate Council of Australia, 2014).

9. S. Connor, "Shock as Retreat of Arctic Sea Ice Releases Deadly Greenhouse Gas," *Independent (London)*, December 21, 2011.

10. A. Fritz, "Massive Hole at End of the Earth Likely Caused by Methane Gas Release," *Washington Post*, August 1, 2014, www.washingtonpost.com/blogs/capital-weather-gang/wp/2014/08/01/massive-hole-at-end-of-the-earth-likely-caused-by-methane-gas-release/ (accessed August 2, 2014).

11. B. Stallard, "'We're f'ed': Methane Plumes Seep from Frozen Ocean Floors," *Nature World News*, August 5, 2014, www.natureworldnews.com/articles/8401/20140805/fd-methane-plumes-seep-frozen-ocean-floors.htm (accessed August 8, 2014).

12. Rio+20 United Nations Conference on Sustainable Development, www.uncsd2012.org (accessed August 20, 2012).

13. J. Hansen et al., "Target Atmospheric CO_2: Where Should Humanity Aim?" *Open Atmospheric Science Journal* 2 (2008): 217–31.

14. M. T. Klare, *The Race for What's Left: The Global Scramble for the World's Last Resources* (New York: Metropolitan Books, 2012); C. A. S. Hall, R. Powers, and W. Schoenberg, in *Biofuels, Solar and Wind as Renewable Energy Systems,* ed. D. Pimentel (Berlin: Singer, 2008), 109–32.

15. J. Cribb, *The Coming Famine: The Global Food Crisis and What We Can Do to Avoid It* (Berkeley: University of California Press, 2010).

16. D. Vaccari, "Phosphorus: A Looming Crisis," *Scientific American Magazine* 300 (2009): 54–59; E. Dolan, "Doomsday: Will Peak Phosphate Get Us Before Global Warming?" *OilPrice.com*, July 22, 2013, http://oilprice.com/Metals/Foodstuffs/Doomsday-Will-Peak-Phosphate-Get-us-Before-Global-Warming.html (accessed August 4, 2014).

17. "Svalbard Global Seed Vault," Government.no, www.regjeringen.no/en/dep/lmd/campain/svalbard-global-seed-vault.html?id=462220 (accessed August 20, 2012).

18. J. Walck and K. Dixon, "Time to Future-Proof Plants in Storage," *Nature* 462 (2009): 721.

19. D. C. Pirages, "Nature, Disease, and Globalization: An Evolutionary Perspective," *International Studies Review* 9 (2007): 616–28.

20. "The Montreal Protocol on Substances That Deplete the Ozone Layer," United Nations Environment Programme, http://ozone.unep.org/new_site/en/montreal_protocol.php (accessed August 4, 2014).

21. Bradshaw and Brook, "Human Population Reduction Is Not a Quick Fix for Environmental Problems"; S. Singh, G. Sedgh, and R. Hussain, "Unintended Pregnancy: Worldwide Levels, Trends, and Outcomes," *Studies in Family Planning* 41 (2010): 241–50.

22. P. R. Ehrlich and A. H. Ehrlich, "The Culture Gap and Its Needed Closures," *International Journal of Environmental Studies* 67 (2010): 481–92.

23. Natural Capital Project, www.naturalcapitalproject.org (accessed August 20, 2012).

24. Millennium Alliance for Humanity and the Biosphere (MAHB), mahb.stanford.edu (accessed August 20, 2012).

25. OccupyWallStreet, occupywallst.org (accessed August 20, 2012).

26. 350.org, www.350.org (accessed August 20, 2012).

27. "Stanford to Divest from Coal Companies," *Stanford News*, May 6, 2014, http://news.stanford.edu/news/2014/may/divest-coal-trustees-050714.html (accessed July 28, 2014).

28. J. Liu et al., "Coupled Human and Natural Systems," *Ambio* 36 (2007): 639–49.

29. J. A. Tainter, *The Collapse of Complex Societies* (Cambridge: Cambridge University Press, 1988).

30. J. C. S. Long, "Piecemeal Cuts Won't Add Up to Radical Reductions," *Nature* 478 (2011): 429; D. Stover, "The Myth of Renewable Energy," *Bulletin of the Atomic Scientists*, November 22, 2011, www.thebulletin.org/web-edition/columnists/dawn-stover/the-myth-of-renewable-energy (accessed February 14, 2015).

31. J. Watts, *When a Billion Chinese Jump* (New York: Scribner, 2010).

32. J. Massola, P. Ker, and L. Cox, "Coal Is Good for Humanity,' Says Tony Abbott at Mine Opening," *Sydney Morning Herald*, October 13, 2014, www.smh.com.au/federal-politics/political-news/coal-is-good-for-humanity-says-tony-abbott-at-mine-opening-20141013-115bgs.html (accessed November 12, 2014).

33. Hansen et al., "Target Atmospheric CO_2."

34. S. Solomon, G.-K. Plattner, R. Knutti, and P. Friedlingstein, "Irreversible Climate Change Due to Carbon Dioxide Emissions," *Proceedings of the National Academy of Sciences of the USA* 106 (2009): 1704–9.

35. Klare, *The Race for What's Left*.

36. J. Hansen, M. Sato, and R. Ruedy, "Perception of Climate Change," *Proceedings of the National Academy of Sciences of the USA* 109 (2012): E2415–23.

37. S. L. Pimm et al., "The Biodiversity of Species and Their Rates of Extinction, Distribution, and Protection," *Science* 344, doi:10.1126/science.1246752 (2014).

38. P. J. Crutzen, "Geology of Mankind: The Anthropocene," *Nature* 415 (2002): 23.

39. Food and Agriculture Organization of the United Nations, *The State of Food Insecurity in the World: Addressing Food Insecurity in Protracted Crises* (Food and Agriculture Organization of the United Nations, Rome, 2010), http://www.fao.org/docrep/013/i1683e/i1683e.pdf (accessed February 13, 2015).

40. G. C. Daily and P. R. Ehrlich, "Impacts of Development and Global Change on the Epidemiological Environment," *Environment and Development Economics* 1 (1996): 309–44.

41. G. C. Daily and P. R. Ehrlich, "Global Change and Human Susceptibility to Disease," *Annual Review of Energy and the Environment* 21 (1996): 125–44.

42. T. Yao et al., "Different Glacier Status with Atmospheric Circulations in Tibetan Plateau and Surroundings," *Nature Climate Change* 2 (2012): 663–67.

43. J McAdam and K. Murphy, "Punishment Not Protection Behind Morrison's Refugee Law Changes," *The Conversation*, June 27, 2014, http://theconversation.com/punishment-not-protection-behind-morrisons-refugee-law-changes-28512 (accessed August 2, 2014).

44. D. B. Lobell et al., "Prioritizing Climate Change Adaptation Needs for Food Security in 2030," *Science* 319 (2008): 607–10; D. D. Lobell and C. B. Field, "Global Scale Climate-Crop Yield Relationships and the Impacts of Recent Warming," *Environmental Research Letters* 2 (2007): 014002.

45. Toon et al., "Consequences of Regional-Scale Nuclear Conflicts."

46. C. J. A. Bradshaw, X. Giam, and N. S. Sodhi, "Evaluating the Relative Environmental Impact of Countries," *PLoS One* 5 (2010): e10440.

CHAPTER ELEVEN

1. Spirit of the West, sotw.ca (accessed August 20, 2012).

2. P. R. Ehrlich and A. H. Ehrlich, "Can a Collapse of Global Civilization Be Avoided?" *Proceedings of the Royal Society B: Biological Sciences* 280 (2013): 20122845.

3. O. Toon et al., "Consequences of Regional-Scale Nuclear Conflicts," *Science* 315 (2007): 1224–25.

4. J. A. Tainter, *The Collapse of Complex Societies* (Cambridge: Cambridge University Press, 1988).

5. T. Blees, *Prescription for the Planet: The Painless Remedy for Our Energy and Environmental Crises* (Charleston, SC: BookSurge, 2008); B. W. Brook and C. J. A. Bradshaw, "Key Role for Nuclear Energy in Global Biodiversity Conservation," *Conservation Biology* 5 (2015): 702–12; W. H. Hannum, "The Technology of the Integral Fast Reactor and Its Associated Fuel Cycle," *Progress in Nuclear Energy* 31 (1997): 10217.

6. National Academy of Sciences USA, in *Population Summit of the World's Scientific Academies* (Washington, DC: National Academies Press, 1993); Union of Concerned Scientists, *World Scientists' Warning to Humanity* (Cambridge, MA: Union of Concerned Scientists, 1993); H. Brown, *The Challenge of Man's Future* (New York: Viking, 1954); P. R. Ehrlich, A. H. Ehrlich, J. P. Holdren, *Ecoscience: Population, Resources, Environment* (San Francisco: W. H. Freeman and Co., 1977); Millennium Ecosystem Assessment, *Ecosystems and Human Well-Being: Synthesis* (Washington, DC: Island Press, 2005); J. Rockström et al., "Planetary Boundaries: Exploring the Safe Operating Space for Humanity," *Ecology and Society* 14, no. 32 (2009), http://www.ecologyandsociety.org/vol14/iss32/art32/.

7. C. J. A. Bradshaw and B. W. Brook, "Human Population Reduction Is Not a Quick Fix for Environmental Problems," *Proceedings of the National Academy of Sciences of the USA* 111 (2014): 16610–15.

8. C. J. A. Bradshaw, X. Giam, and N. S. Sodhi, "Evaluating the Relative Environmental Impact of Countries," *PLoS One* 5 (2010): e10440.

9. B. Gardiner, "Biodiversity Is the Spark of Life," *BBC News*, May 7, 2009, news.bbc.co.uk/2/hi/science/nature/8034412.stm (accessed August 20, 2012).

10. B. Moyers and M. Winship, "The Iraq Was All about Oil, All Along," *Alternet*, July 5, 2008, www.alternet.org/story/90509/the_iraq_war_was_about_oil,_all_along (accessed August 7, 2014).

11. "Preventing and Controlling Micronutrient Deficiencies in Populations Affected by an Emergency," Joint Statement by WHO, World Food Programme, & UNICEF, www.unicef.org/nutrition/files/Joint_Statement_Micronutrients_March_2006.pdf (accessed August 4, 2014);

12. J. E. Stiglitz, "Of the 1%, by the 1%, for the 1%," *Vanity Fair*, May 2011, www.vanityfair.com/news/2011/05/top-one-percent-201105 (accessed February 14, 2015).

13. A. Bartlett, "The Arithmetic of Growth: Methods of Calculation," *Population and Environment: A Journal of Interdisciplinary Studies* 14 (1993): 359–87.

14. M. Huesemann and J. Huesemann, *Techno-Fix: Why Technology Won't Save Us or the Environment* (Gabriola Island, British Columbia, Canada: New Society Publishers, 2012).

15. H. Daly, "Growth, Debt, and the World Bank," *Ecological Economics* 72 (2011): 5–8.

16. W. L. Robinson, *Global Capitalism and the Crisis of Humanity* (New York: Cambridge University Press, 2014).

17. P. R. Ehrlich, "A Personal View: Environmental Education—Its Content and Delivery," *Journal of Environmental Studies and Sciences* 1 (2011): 6–13.

18. Tainter, *The Collapse of Complex Societies.*

19. K. Arrow et al., "Economic Growth, Carrying Capacity, and the Environment," *Science* 268 (1995): 520–21; P. Dasgupta, *Human Well-Being and the Natural Environment* (Oxford: Oxford University Press, 2001).

20. D. Lam, "How the World Survived the population Bomb: Lessons from 50 Years of Extraordinary Demographic History," *Demography* 48 (2011): 1231–62.

21. World Hunger Education Service, "2014 World Hunger and Poverty Facts and Statistics."

22. K. Rawe, *A Life Free from Hunger: Tackling Child Malnutrition* (London: Save the Children UK, 2012).

23. "Report Sought on India Farm Suicides," *BBC News India*, December 21, 2011, www.bbc.co.uk/news/world-asia-india-16281063 (accessed August 20, 2012).

24. H. Gardiner, "Poor Sanitation in India May Afflict Well-Fed Children with Malnutrition," *New York Times*, July 13, 2014, A1–A6.

25. J. Harte, "Human Population as a Dynamic Factor in Environmental Degradation," *Population and Environment* 28 (2007): 223–36.

26. A. D. Barnosky et al., "Approaching a State Shift in Earth's Biosphere," *Nature* 486 (2012): 52–58.

27. P. R. Ehrlich, "Key Issues for Attention from Ecological Economists," *Environment and Development Economics* 13 (2008): 1–20.

28. M. Spence, *The Next Convergence: The Future of Economic Growth in a Multispeed World* (New York: Farrar, Straus, and Giroux, 2011).

29. K. E. Boulding, In *Environmental Quality in a Growing Economy*, ed. H. Jarrett (Baltimore: Johns Hopkin University Press, 1966), 3–14.

30. Rockström et al., "Planetary Boundaries."

31. R. L. Naylor, ed., *The Evolving Sphere of Food Security* (Oxford: Oxford University Press, 2014).

32. H. Welzer, *Climate Wars: What People Will Be Killed for in the 21st Century* (Oxford: Polity, 2012).

33. W. I. Robinson, *Promoting Polyarchy: Globalization, US Intervention, and Hegemony* (Cambridge: Cambridge University Press, 1996).

34. "As Bush Ship Continues to Sink, Alan Greenspan Claims Iraq War Was Really for Oil," *Natural Law Publishing*, September 16, 2007, www.nlpwessex.org/docs/greenspaniraq.htm (accessed August 4, 2014).

35. I. R. Hancock, "Holt, Harold Edward (1908–1967)," *Australian Dictionary of Biography* (1996), http://adb.anu.edu.au/biography/holt-harold-edward-10530 (accessed August 4, 2014).

36. "Ex-Harper Appointee Calls Canada a 'Rogue State' on Environment," *CTV News*, December 2, 2013, www.ctvnews.ca/canada/ex-harper-appointee-calls-canada-a-rogue-state-on-environment-1.1570676 (accessed August 4, 2014).

37. "Medieval Canada Threatens Global Biodiversity," *ConservationBytes.com*, November 25, 2013, http://conservationbytes.com/2013/11/25/medieval-canada-threatens-global-biodiversity (accessed August 4, 2014).

38. P. Beinart, "Dick Cheney Just Buried the Bush Doctrine," *The Atlantic*, June 27, 2014, www.theatlantic.com/international/archive/2014/06/dick-cheney-just-buried-the-bush-doctrine/373621/ (accessed August 4, 2014).

39. Toon et al., "Consequences of Regional-Scale Nuclear Conflicts"; P. R. Ehrlich et al., "Long-Term Biological Consequences of Nuclear War," *Science* 222 (1983): 1293–300.

40. E. Rosenberg, "Experts Warn of an Accidental Atomic War," *SFGate*, October 6, 2006, www.sfgate.com/news/article/Experts-warn-of-an-accidental-atomic-war-2550308.php (accessed August 4, 2014).

41. G. Brumfiel, "Megatons to Megawatts: Russian Warheads Fuel US Power Plants," *NPR*, December 11, 2013, www.npr.org/2013/12/11/250007526/megatons-to-megawatts-russian-warheads-fuel-u-s-power-plants (accessed August 4, 2014).

42. T. Woodroofe, "Australia Must Walk the Walk on Nuclear Weapons," *ABC*, October 10, 2013, www.abc.net.au/news/2013-10-11/woodroofe-australia-must-walk-the-walk-on-nuclear-weapons/5016626 (accessed August 4, 2014).

43. Toon et al., "Consequences of Regional-Scale Nuclear Conflicts."

44. D. J. Davidson, J. Andrews, and D. Pauly, "The Effort Factor: Evaluating the Increasing Marginal Impact of Resource Extraction Over Time," *Global Environmental Change* 25 (2014): 63–68; D. J. Davidson and J. Andrews, "Not All about Consumption," *Science* 339 (2013): 1286–87.

45. H. Vatsikoupouls, "Australia Has More Soft Power Than Ever but Can We Keep It?" *The Conversation*, December 4, 2013, http://theconversation.com/australia-has-more-soft-power-than-ever-but-can-we-keep-it-20698 (accessed August 4, 2014).

46. Blees, *Prescription for the Planet*; Hannum, "The Technology of the Integral Fast Reactor and Its Associated Fuel Cycle"; S. Hong, C. J. A. Bradshaw, and B. W. Brook, "Nuclear Power Can Reduce Emissions and Maintain a Strong Economy: Rating Australia's Optimal Future Electricity-Generation Mix by Technologies and Policies," *Applied Energy* 136 (2014): 712–25.

47. I. McGregor, "Australia's Coal Industrty Needs to Prepare for Global Climate Action," *Business Spectator*, July 3, 2014, www.businessspectator.com.au/article/2014/7/3/policy-politics/australias-coal-industry-needs-prepare-global-climate-action (accessed August 4, 2014).

48. A. Cameron, "Mr. Abbott, Keep God Out of Politics," *The Age*, March 16, 2014, www.theage.com.au/comment/mr-abbott-keep-god-out-of-politics-20140315-34u51.html (accessed August 4, 2014).

49. "National Human Rights Action Plan," *Human Rights Law Centre*, www.humanrightsactionplan.org.au/nhrap/focus-area/womens-rights (accessed August 4, 2014).

50. "What Contraceptive Are You?," www.whatcontraceptiveareyou.com.au (accessed August 4, 2014); "Fact Sheet: Uninteded Pregnancy in the United States,"

Guttmacher Institute, www.guttmacher.org/pubs/FB-Unintended-Pregnancy-US .html (accessed August 4, 2014).

51. "Australian Abortion Law and Practice," Children by Choice, www.children bychoice.org.au/info-a-resources/facts-and-figures/australian-abortion-law -and-practice (accessed August 4, 2014).

52. "States Continue to Enact Abortion Restrictions in First Half of 2014," Guttmacher Institute, www.icontact-archive.com/s5XvpXoL5HYHElU3Q3k_wkayKXq _x0sx?w=4 (accessed August 4, 2014).

53. J. Price, "Christoper Pyne wins Ernie Award 2014 for His Comments on the Impact of Uni Fees on Women," *Daily Life*, September 25, 2014, www.dailylife.com .au/news-and-views/news-features/christoper-pyne-wins-ernie-award-2014 -for-his-comments-on-the-impact-of-uni-fees-on-women-20140925-3gjsu.html (accessed September 25, 2014).

54. N. S. Sodhi, P. Davidar, and M. Rao, "Empowering Women Facilitates Conservation," *Conservation Biology* 143 (2010): 1035–36.

55. J. B. Casterline and S. W. Sinding, "Unmet Need for Contraception in Developing Countries and Implications for Population Policy," *Population and Development Review* 26 (2000): 691–723.

56. K. Arrow et al., "Are We Consuming Too Much?" *Journal of Economic Perspectives* 18 (2004): 147–72.

57. H. E. Daly and K. N. Townsend, eds., *Valuing the Earth: Economics, Ecology and Ethics* (Cambridge, MA: MIT Press, 1993).

58. Bradshaw and Brook, "Human Population Reduction Is Not a Quick Fix for Environmental Problems."

59. P. R. Ehrlich and A. H. Ehrlich, "The Culture Gap and Its Needed Closures," *International Journal of Environmental Studies* 67 (2010): 481–92.

60. S. Solomon, G.-K. Plattner, R. Knutti, and P. Friedlingstein, "Irreversible Climate Change Due to Carbon Dioxide Emissions," *Proceedings of the National Academy of Sciences of the USA* 106 (2009): 1704–9; J. Hansen et al., "Target Atmospheric CO_2: Where Should Humanity Aim?" *Open Atmospheric Science Journal* 2 (2008): 217–31.

61. J. A. Foley et al., "Solutions for a Cultivated Planet," *Nature* 478 (2011): 332–42.

62. A. J. McMichael, J. W. Powles, C. D. Butler, and R. Uauy, "Food, Livestock Production, Energy, Climate Change, and Health," *The Lancet* 370 (2007): 1253–63.

63. D. Shindell et al., "Simultaneously Mitigating Near-Term Climate Change and Improving Human Health and Food Security," *Science* 335 (2012): 183–89.

64. S. Hong, C. J. A. Bradshaw, and B. W. Brook, "Nuclear Power Can Reduce Emissions and Maintain a Strong Economy: Rating Australia's Optimal Future Electricity-Generation Mix by Technologies and Policies," *Applied Energy* 136 (2014): 712–25.

65. P. R. Ehrlich and R. E. Ornstein, *Humanity on a Tightrope: Thoughts on Empathy, Family, and Big Changes for a Viable Future* (New York: Rowman & Littlefield, 2010).

66. S. Herrando-Pérez, S. Delean, B. W. Brook, and C. J. A. Bradshaw, "Density Dependence: An Ecological Tower of Babel," *Oecologia* 170 (2012): 585–603.

67. Brook and Bradshaw, "Key Role for Nuclear Energy in Global Biodiversity Conservation."

68. J. Harte and M. E. Harte, *Cool the Earth, Save the Economy: Solving the Climate Crisis Is Easy* (2008), http://cooltheearth.us.

69. B. Heard, C. J. A. Bradshaw, and B. W. Brook, "Beyond Wind: Furthering Developing of Clean Energy in South Australia," *Transactions of the Royal Society of South Australia* 139 (2015): 57–82.

70. Brook and C. J. A. Bradshaw, "Key Role for Nuclear Energy in Global Biodiversity Conservation."

71. Hong, Bradshaw, and Brook, "Nuclear Power Can Reduce Emissions and Maintain a Strong Economy"; Heard, Bradshaw, and Brook, "Beyond Wind."

72. Brook and Bradshaw, "Key Role for Nuclear Energy in Global Biodiversity Conservation."

73. A. M. Weinberg, "Technological Optimism," *Society* 17 (1980): 17–18.

74. M. T. Klare, *The Race for What's Left: The Global Scramble for the World's Last Resources* (New York: Metropolitan Books, 2012); H. Knaup and J. von Mittelstaedt, "The New Colonialism: Foreign Investors Snap Up African Farmland," *Der Spiegel*, July 30, 2009, http://www.spiegel.de/international/world/0,1518,639224,00.html (accessed March 3, 2015).

75. "Foreign Ownership of Australian Farmland," *Crikey*, March 8, 2015, http://www.crikey.com.au/foreign-ownership-of-farmland (accessed March 16, 2015).

76. Extreme Citizen Science (EXCITES), www.ucl.ac.uk/excites (accessed August 20, 2012).

77. 350.org, www.350.org (accessed August 20, 2012).

78. Growth Busters, www.growthbusters.org (accessed August 20, 2012); Call of Life: Facing the Mass Extinction, calloflife.org (accessed August 20, 2012).

79. *Solutions: For a Sustainable and Desirable Future*, www.thesolutionsjournal.org (accessed August 20, 2012).

80. The Daily Climate, www.dailyclimate.org (accessed August 20, 2012).

81. Millennium Alliance for Humanity and the Biosphere (MAHB), mahb.stanford.edu (accessed August 20, 2012).

82. GetUp! Action for Australia, www.getup.org.au (accessed August 4, 2014).

83. J. Chait, "The Right's '53 Percent' Solution to Occupy Wall Street," *New York Magazine*, October 11, 2011, nymag.com/daily/intel/2011/10/the_right_answers_occupy_wall.html (accessed August 20, 2012).

84. R. Reich, "The Seven Biggest Economic Lies," *Reader Supported News*, October 12, 2011, www.readersupportednews.org/opinion2/277-75/7845-the-seven-biggest-economic-lies (accessed August 20, 2012).

85. A. W. Blumrosen and R. G. Blumrosen, *Slave Nation: How Slavery United the Colonies and Sparked the American Revolution* (Naperville, IL: Sourcebooks, 2006).

86. "Federalist Project 15," July 23, 2011, liberaccademia.wordpress.com/2011/07/23/federalist-project-15 (accessed August 20, 2012).

87. P. Lewis, "Australia's Economy Is Healthy, So How Can There Be a Budget Crisis?" *The Conversation*, May 5, 2014, http://theconversation.com/australias

-economy-is-healthy-so-how-can-there-be-a-budget-crisis-26036 (accessed August 2, 2014).

88. L. Minchin and M. Hopkin, "Carbon Tax Axed: How It Affects You, Australia and Our Emissions," *The Conversation*, July 15, 2014, http://theconversation.com/carbon-tax-axed-how-it-affects-you-australia-and-our-emissions-28895 (accessed August 4, 2014).

89. Robinson, *Global Capitalism and the Crisis of Humanity*.

90. Ibid.

91. C. Rootes, "Environmentalism in Australia," *Environmental Politics* 10 (2001): 134–39.

92. B. Tranter, "Environmentalism and Education in Australia," *Environmental Politics* 6 (1997): 123–43.

93. Ehrlich, "A Personal View."

94. Ehrlich and Ornstein, *Humanity on a Tightrope*.

95. D. Lam, "How the World Survived the population Bomb: Lessons from 50 Years of Extraordinary Demographic History," *Demography* 48 (2011): 1231–62.

96. Bradshaw, Giam, and Sodhi, "Evaluating the Relative Environmental Impact of Countries"; W. E. Rees, in *Encyclopedia of Biodiversity*, vol. 2, ed. S. A. Levin (San Diego, CA: Academic Press, 2001), 229–44.

97. I. Kubiszewski et al., "Beyond GDP: Measuring and Achieving Global Genuine Progress," *Ecological Economics* 93 (2013): 57–68.

98. P. Kareiva, H. Tallis, T. H. Ricketts, G. C. Daily, and S. Polasky, eds., *Natural Capital: Theory and Practice of Mapping Ecosystem Services* (Oxford: Oxford University Press, 2011).

99. D. Diderot, In *The Macmillan Dictionary of Quotations*, ed. J. Daintith (Edison, NJ: Chartwell Books, 2000), 34.

100. D. Adams, *The Restaurant at the End of the Universe* (London: Pan Books, 1980).

101. S. Lawler, "Scientists Wary of Wearing Their Hearts on Their Sleeves—Is This What We Really Want?" *The Conversation*, August 23, 2012, theconversation.edu.au/scientists-wary-of-wearing-their-hearts-on-their-sleeves-is-this-what-we-really-want-9016 (accessed August 24, 2012).

102. "Canadian Scientists Take to the Streets," *Feminist Philosophers*, July 11, 2012, http://feministphilosophers.wordpress.com/2012/07/11/canadian-scientists-take-to-the-streets/ (accessed August 4, 2014).

103. "Conservation Value of Paddy Wagon Currency: Civil Disobedience by Scientists," *ConservationBytes.com*, May 12, 2012, http://conservationbytes.com/2012/05/12/conservation-value-of-paddy-wagon-currency/ (accessed August 4, 2014).